WINGS OF THE DAWN

THE LIFE OF CYRIL JOHNSON

A BIOGRAPHY

BY KATE JOHNSON

Wings of the Dawn
The Life of Cyril Johnson

Published on behalf of the author by
dpaPublishing

Please note: All mention of distances throughout this book are in the Imperial measurement system in use during the period. The following conversion table will be of some assistance to those unfamiliar with the former system.

100 feet converts to 30.48 metres
500 feet converts to 152.4 metres
1000 feet converts to 304.8 metres
100 yards converts to 91.44 metres
1000 yards converts to 914.4 metres
1 mile converts to 1.6 kilometres
1 kilometre converts to 1093.6 yards
500lbs converts to 226kg
1000lbs converts to 553kg

ISBN: 978-1-921207-16-7

Designed, printed and bound at
Digital Print Australia
135 Gilles Street Adelaide South Australia 5000
www.digitalprintaustralia.com

Dedication

To the next generation, those with or without military or aviation backgrounds.

May this ignite curiosity in the events that have shaped your world and the personal stories of other unsung heroes, and with any luck, may it inspire you to record their incredible stories too.

Contents

Preface

A few words from Cyril's second son, Steve Johnson.

My father has always been my hero and when you read this book you will see why.

Whilst this book is a fascinating read, it is important because of the profound message that it reveals.

I have had and continue to lead a very adventurous and at times bizarre life that has taken me to most corners of the world. When combined with my career in heavy engineering and my interests in flying vintage aircraft, I have been exposed to some of the most extraordinary people imaginable. That said, it may surprise some to know that although I am now in my late 50's, my father is still my hero, and from my perspective commands the same respect as would lifelong heroes such as Nelson Mandela and Neil Armstrong. Perhaps in part the fascination is because his achievements have been so understated.

I think I was aware of a good proportion of the facts and accounts contained in this book but only as disjointed and incomplete, though very interesting, accounts of various events at different points in his life. Now having read the consolidated story from beginning to end, I confess to being stunned by the implications of the story that is presented in such a matter of fact style about a life of achievement that defies all the odds. It is indeed an example of Aristotle's concept of "the whole is greater than the sum of its parts".

Cyril is a man of so many contradictions; a man of incredible bravery and resolve but so inconspicuous. He does not suffer fools but is not one to force lengthy dissertations and free advice on the unreceptive.

Cyril Johnson, nearing his 94th birthday

When he does have something to say it is generally worth listening to.

In my case as I graduated as a civil engineer he gave me three short succinct and typically understated pieces of advice that was to profoundly influence my career, my life and those around me in ways I could not then have imagined. He said to me:

- *Don't pick a fight unless you are prepared to go all the way;*

- *If you believe it be prepared to say it even if you are the only one standing;* and
- *No matter the personal cost and pain never compromise your integrity.*

He of course had no way of knowing how this would be put to the extreme test at various points in my life—or did he?

As a child I saw what pain could do and remember vividly and with fear those times when he would actually pass out from the pain of his back injuries and malaria relapses, yet it did not defeat him and his duty to his family. My childhood was a magical adventure for myself, my brother and my sister because of the strength, stability and safety he provided which is still evident today as he is in his 90's. The depth and quality of my mother and father's marriage was such that so many people from so many walks of life would see them as second parents.

My father is a walking medical mystery, having suffered every possible reason not to go to war in the first place, and then having gone to war, he had every possible reason to opt out.

At the start of World War II he tried to enlist but was proclaimed to be permanently unfit to join the services because of a damaged heart valve from a childhood illness. Undeterred he found a way around this small obstacle and joined the Royal Air Force. This could have resulted in a court martial had his records ever caught up with him. When on active flying duties at the centre of some of the most appalling circumstances, he could have, and should have, been invalided out of the Air Force having contracted malaria whilst flying in North Africa or because of his back injuries caused by crash landing a Blenheim Bomber on the beach in Pakistan.

At 94 years of age Cyril can be quite frustrated at his inability to do all the things he wants to do. However he still walks with little or no assistance; can drive himself; and lives independently. If you enter into a conversation on almost any subject, he will hold his own for hours. He can still recite classic poetry for half an hour or solve an algebraic quadratic equation in his head.

A few years ago I took him to the Canberra War Museum as a birthday present. He stood under G for George, the Avro Lancaster Bomber, like the ones he flew on night raids over Germany and just whispered to himself —"*I was only 24*" and then he stood there for what was for me one of the most pregnant pauses I have known and I saw tears in his eyes. He did not need to and did not say any more, I just saw it all in his eyes and that moment still sends shivers up my spine and brings tears to my eyes.

Why is this book important?

As a 'baby boomer' approaching retirement, I became increasingly dismayed about the ticking bomb that we will leave to our children and grandchildren. The physical mess resulting from an over exploited planet of finite resources is merely a symptom of the real mess that threatens the essence of human existence and values.

Whether we choose to learn from or ignore the past will determine how the future of mankind plays out and yet the best we seem to do, and we don't do that at all well, is to turn our attention to the physical tangible problems of today. Those problems will only grow at an exponential rate if we continue to ignore the human lessons of the past and as a result leave our children to blunder through unimaginable consequences of conflicts yet to come. Such conflicts are unnecessary but are almost inevitable if we do not learn from the last 100 years.

There are very few people left who have profoundly experienced the realities of getting it wrong on a global scale over the last century. Of those, there is a minuscule number who can, with spine chilling clarity, allow us to live through their eyes and see with real insight the biggest changes in the history of human existence, all of which has occurred over the last 100 years and so much of which already seems to have been forgotten or perhaps its profound impact upon the future of humans has never really been understood.

We tend to dismiss such recollections of an older generation as curious and interesting historical facts but without any real connection to ourselves and we go about our daily worries perhaps commenting on where we see the world is heading, whilst not recognising that nothing much is new when it comes to human behaviour. So we are oblivious to the lessons than must be learnt if we are not going to repeat those mistakes with even greater physical consequences.

The importance of this book is that it exposes profound insights through the eyes of a person who has not only lived through but in so many ways experienced and been part of the extraordinary times of pre-war, world-war and post-war life, as a person of extraordinarily strong character but also as a person of no great fame, power, wealth, notoriety or high profile.

This book presents truly compelling and chilling insights to the realities of the best and worst aspects of humankind over the last century.

So why is he still my hero?

Today we seem to idolise rich people, sports people, movie stars, music stars and politicians because of the role they occupy through circumstance and talent but rarely because of their true underlying character. These are mostly artificial creations in that they act or they are the creations of image makers or that they are prominent because of the power or money they control.

Without these artificial props one must ask: would they still be considered as heroes?

My father is none of these but rather he is a humble, deep-thinking person who has faced unimaginable fears with simple courage and determination to do what was right. He did what he had to do whilst always carefully weighing the rights and wrongs of doing so and he did it for no personal gain or fame.

My daughter Kate, who herself has led an amazing and adventurous life so far, has managed over the past years to patiently extract the substance of this book to an extent that no one else could have done. Were it not for my daughter's persistence you would not get to know of this remarkable person. He never had, never needed and never sought any recognition because he is by nature 'the unsung hero'.

Foreword

Kate Johnson, granddaughter of Cyril Johnson and author of his biography.
Photo courtesy Silver Photography, Mel Alexander

THROUGHOUT HIS LIFE, CYRIL DID not have the opportunity to know his father, Captain James Johnson, particularly well, as his father was often away at sea, and even when he was not, through the busy young life of Cyril, and the Captain's stiff upper lip, much of his father's story had not been passed on. Particularly as a young man in his twenties and thirties, Cyril supposed he had navigated his own path. However, many years later, when he retired in the 1980's, he was fortunate to have the opportunity to read his father's large collection of diary entries, letters and ships' logs, revealing his innermost thoughts, and it was then that Cyril discovered just how similar their life journey, perspectives, characteristics and values really were.

Based on this life-changing read, and an old promise that Cyril had made to his mother's sister, Aunt Teresa, at the beginning of the Second World War, he set about transcribing some of his own diary entries in the late 1980's. By the early 1990's Cyril had transcribed the majority of the diary entries that are included in this book, as he felt that he owed it to his children and grandchildren to make them available, 'should anybody be interested'.

Cyril then typed some of the stories from his early years, to give his diary entries and love of flying a context, in what would form the outline of the beginning of this book. One Christmas, in the mid 1990's, Cyril presented his typed notes to his three adult children.

My father, uncle and aunt each received a thin booklet, which had been proudly printed by my grandfather on his dot matrix printer, with the edge strips torn off to make way for the binder. I was still in primary school, yet I clearly remember reading the first pages of his gift and being transported into a completely different world.

As a child I was excited to learn about my grandfather's history, and would frequently ask him to elaborate on stories from when he was my age. We would talk about the types of games he played when he was at school, his pets, the sorts of cars they drove, but never much about the war. I always knew there was more to this fun and witty grandfather, but other than knowing that his poor hearing and back operations were due to 'the war', I had little idea of just how amazing his story was. As children tend to be, I was curious and tried to imagine what the 'olden days' were like, and once, after looking at his photo albums, I asked if the world used to be black and white —for which I promptly received a lesson on the limitations of photography during that period!

As a family, we were incredibly proud to see what Cyril had been working on, and encouraged him to write more of his story. Over the years that followed, various family members would frequently suggest that he continue with his project and write a book, however, in a style true to his frequently understated nature, his standard reply became, 'but it's just an ordinary life'. Fortunately, we had grown up in a close family, so we knew that he had a wealth of fascinating stories, crystal clear memories and that despite his modesty, he lived an inspirational life.

When I was fourteen years old, I took flying lessons, and not being old enough to have a driver's licence, my father and grandfather would drive me to the flying school. Not yet knowing most of my grandfather's story, I was amazed by his ability to instinctively understand the various aircraft that I was learning to fly on. He has always retained his ability to look inside a cockpit and work everything out from first principles. After one lesson in particular, I remember him sharing with me a single line of advice—'*always know the limits of your machine, and particularly the limits of yourself*'—wisdom which I have been able to apply in many other contexts.

Well over a decade later, after Cyril wrote some of the memories included in these early chapters, and the project had stalled; I asked my grandfather, if I could have the honor of continuing to write his life story. Again, he didn't think there was much of a story worth telling, but I persisted. What formed was an incredible tradition in which the two of us would devote many weeks a year, during most of my university holidays, working on this book. The project was incredibly fascinating and enjoyable, and the more we tapped into his memories, the more I couldn't wait to return for the next moving and exciting instalments.

We developed an interviewing process, in which we would record Cyril telling his stories, starting at the beginning, in 1920, and each time I asked Cyril to explain one story, two more would emerge. His recollections were captivating and the book took on a life of its own. When I first started interviewing him, I had very little idea of the overall story line, and there were frequent sessions during which I would cancel my other plans for the afternoon or evening, so captivated by his stories that I preferred instead to stay on a few more hours listening to what would unfold.

The more we progressed with the storyline, the more I was astonished. Although I must admit, as with most treasures, the jewels were often hidden, and he would happily let us pass them by. I soon cottoned onto his humble style and learnt to keep revisiting old stories, having to really dig for the gold. Eventually, in bursts of cheeky laughter or subtle tears, and often a mixture of both, the details would emerge, and I would be left with a story that I could not wait to write up in its complete form, with the full thrust of its emotion, heroism and bravery. Over the years, I would frequently revisit old anecdotal stories, and discover that they were in fact some of the most profound moments he had experienced, and as I expect is the case with a lot of great achievers, because he had no need for accolades, this meant I had to learn to pick up on the little throw away lines; they were only the tip of the iceberg.

The ways in which this book came together highlighted to me a small sample of the incredible change that Cyril had lived through. Many of the stories and quotes came from a box of worn leather bound diaries, with crisp yellowed pages. Cyril had since typed most of the diary entries, and saved them on a 5¼ inch floppy disk, in an effort to preserve them in a digital format that would last. Such files were eventually updated and copied onto a CD, before being stored on a USB and ultimately backed up somewhere in 'The Cloud'. The vast majority of the events in this book were captured through the use of an inbuilt recorder on my smart phone. I would interview Cyril, and record his narratives, often for several hours at a time, before going home to type them out on a laptop that was thinner than most of his notebooks, and yet to his full credit, he was able to keep up with all the changing formats, and he sent emails as appropriate for research into his family history and photos. It was an amazing insight into the sorts of changes he must have lived through and even in his nineties, is still able to embrace.

The generational gaps did not stop there; I found many of the terms and references that he used, once common place, I had never heard before. Cyril had already started parts of the book, and given the way in which we have managed to capture so many of his stories, as he tells them, I decided to narrate the story in his voice, telling it in the first person. I found that this worked well because it enabled the written text to reflect his views as closely as possible, and capture his understated and optimistic nature. To attempt to explain his thoughts and views in any other words would have lost something of the spritely ways in which he recalls even the most difficult and painful memories.

For this reason I have taken great pains to ensure that the book was written as closely as possible to his normal language, retaining the vast array of classic English expressions which he uses, whilst integrating the necessary background information, being mindful that the goal has always been to make his story accessible to future generations, and that sadly, World War II, is no longer covered well in school history lessons, if at all. Ultimately, Cyril and I hope that this is a book which captivates people of all ages, those with or without aviation or military backgrounds, to understand and appreciate what it felt like to live through the times of such unprecedented change.

As our modern lives get busier and busier, it is important to remember that there are fewer and fewer people around today with such incredible stories. I see it as a privilege to be able to capture his stories in a way that can be passed down in a modern and relevant form to generations to come. By embarking on this venture, I have grown and been more blessed than I ever imagined. It has been without a doubt one of the most fulfilling experiences to be invited into his incredible world, and be able to make something lasting for others to appreciate.

Whilst working on this project, many have remarked on the rarity of a woman in her twenties making the time to write about the life of a veteran who was born in the 1920's, but it need not be a rare thing at all. I would encourage everybody with loved ones who have lived through such incredible times to do the same. The lessons learnt from the lives already lived are important if history is to avoid repeating itself. To anyone who is interested in the themes of this book, I suggest you start your own process of interviewing a loved one. It may be challenging to get the stories flowing at first, but once the well is open, you will be amazed at the depth of wisdom and experiences that flow. Don't wait until you have the time, make the time, for your sake, and that of your children and grandchildren, who I am sure will thank you!

Acknowledgments

A heartfelt thank you goes to:

Cyril's sons, Alastair and Steve for prompting me to ask questions about specific stories, particularly those in which Cyril is a real hero, where the praises due to him would otherwise have been overlooked due to Cyril's humble nature.

In particular, a thank you to Cyril's eldest son for the countless hours of kind deeds and unseen services that go towards supporting Cyril and enabling him to live the independent life that he still enjoys. Thank you also to Cyril's second son for reading drafts and assisting with various technical aviation aspects and for believing so passionately in this book.

Cyril's grandson, my brother, for his endless reinforcement, particularly with all his technical and computer support, and encouraging reminders of how important this work is for our generation and those to follow. Thank you also for producing the website at **www.wingsofthedawn.com.au**

Indebted gratitude goes towards those who made this concept a reality; particularly, by volunteering to read drafts, providing insight from a wide variety of perspectives, interests, ages and professional backgrounds, and assisting with everything from technical facts, historic backgrounds, proof reading and suggestions. We are truly grateful for the enormous number of hours you have given to this work.

In particular, thank you to my mother, who has always supported me in all of my endeavours, and has invested untold hours into this book. You have always taught me that no matter how busy you are, there are always more hours in the day when it comes to helping family.

Enormous recognition goes to my wonderful husband,

who has protected me from the demands of the outside world and supported me in every possible way during the time that it took to complete the book.

Thank you Cyril Johnson for letting us share in your incredible insight and world views as you recalled and explained your memories and experiences. We are so grateful to have a record of your inspirational life in a form that can be passed on to future generations. Let us also be clear that writing this book was your inspiration, so may all complaints be directed accordingly!

Finally, thank you to the family and friends who first encouraged Cyril to write a book, and then, after his 90th birthday, redirected their efforts towards encouraging me to write Cyril's book.

We trust you enjoy this work in its completed form, at last! Please, if you have any friends and family members who 'should write a book some day', encourage them to do it while they are still young, or you may be the one left with the honour of doing this!

Chapter One
1920

I WAS BORN IN THE small town of Ormskirk, Lancashire, England, in January 1920, the youngest of a family of five children. My father, Captain James Johnson, was discharged from the Royal Navy on 31st December 1919, a mere 16 days before my arrival and returned to sea as a merchant immediately after my birth. He had served as a transport staff officer at Le Havre, France until 9th October 1919 when his commanding officer, Captain Hamilton, wrote of him as "a very energetic and zealous officer". Between then and 31st December 1919, his whereabouts are unclear, but I assume he was on leave in April 1919 or otherwise my story would be a little presumptuous! Indeed, I never heard anything to suggest otherwise, particularly given my mother's virtuous character.

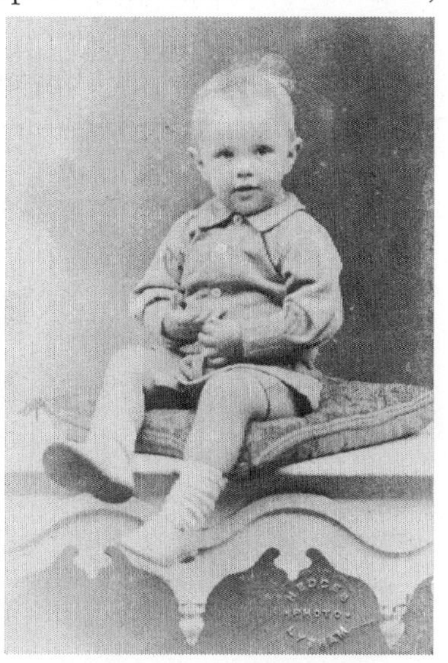

Cyril Alphonsus Bede Johnson, August 1921.

My mother, Aida Monica Mary Cooke, born in 1877, was very ill at the time of my birth and for some time afterwards. I was not a robust child either and considering that, I was taken to church and baptised into the Catholic Church at a very young age.

My mother was too sick to travel, so my uncles and extended family transported me in Grandfather's green Buick EK-78, a limousine model that gave away his prosperous financial status.

It was at my baptism that I was officially named Cyril Alphonsus Bede Johnson, although these names were not likely to have been my parents' choice and they were certainly not mine. Indeed, why anybody would name a young child after an eighth century venerable monk, knowing that they would eventually have to survive boarding school, eludes me but I am sure it was all in good humour. In any case, it seems likely that my father, James, was not present at my baptism either, because if he had been, he would have overridden such suggestions and chosen more traditional names as he had for my four older siblings, Monica, John, Elizabeth, and Frances.

My father's absence at my baptism was certainly no surprise, as he was regularly away at sea for up to three years at a time working in West Africa as the captain of a large merchant sailing ship. Yet despite this difficult lifestyle, he was a family man who loved my mother greatly and not surprisingly, the dates on our family tree represent a pattern similar to his return voyages, with the requisite nine-month lag. Although he was away for so much of the time, he always retained his place as the head of the household and my siblings and I had enormous respect for him.

Father wrote letters and telegrams to each member of the family whenever he had the opportunity and of course we were all very excited every time a message would arrive from 'The Captain'. Almost every month one of my older siblings would run excitedly into the house waving an envelope, calling for everyone to gather around so that we could hear his latest letter being read aloud, usually by my mother. I always loved hearing from father and learnt to treasure his letters and keepsakes enormously, for which I am now very thankful, and no doubt this contributed to my habit of keeping so many other letters, photos, cards, mementoes, and diaries over the past ninety years.

The letter that best captures the softer aspect of his character was written when I was eighteen months old.

Freados.
July 1st 1921.

Dear Little Cyril.
Just a little note for daddy's big boy, although you will not be able to understand it, yet perhaps in the years to come you may like to read your early letters. I hear you are growing into quite a big man, and very fond of teasing. I wonder if you will tease me when I get back home to you, or will you be shy like the others were. Give mum a big big kiss for me and lots of love & kisses to you from

Your loving daddy.
J. Johnson

× ×
× ×

A letter from Cyril's father, Captain James Johnson, on the 1st July 1921.

"Dear little Cyril, Just a little note for Daddy's big boy. Although you will not be able to understand it yet, perhaps in the years to come you may like to read your early letters. I hear you are growing into quite a big man and very fond of teasing. I wonder if you will tease me when I get back home to you or will you be shy like the others were? Give mum a big kiss for me, and lots of love and kisses to you, from your loving daddy. J Johnson"

The Captain's letters to the family captured a glimpse into his soft fatherly nature, an aspect that was rarely, if ever, seen by anybody outside the family. To everybody else James was a stern ship's captain capable of making grown men tremble, a contrast I came to appreciate early in life and particularly during my teenage years. It wasn't as though he was ever aggressive, but more powerfully than that, he ruled with a quiet strength and air of self-confidence that made him almost impossible to cross.

Indeed, I remember hearing that on one of his return

voyages, he took the bus home from Liverpool. A passenger later relayed that 'The Captain' had rung the bell shortly before his stop, but the bus driver was busy chatting to the conductor. The driver drove past the stop and failed to slow down. Father got up and walked down to the front of the bus where, according to the passenger, a few words were exchanged in such a quiet tone that none of the passengers could hear them. Suddenly, the conductor walked towards the back door and took a seat while the driver came to a stop and then reversed the bus for about half a mile to the intended stop! It seemed that everybody respected 'The Captain', and I could see that he would have had no trouble keeping the roughest of sailors in line.

I was always excited to receive letters from 'The Captain', but I can also identify with the poor bus driver and his reaction to my father's intimidating nature. Indeed, my most frightening memory of my young life was the moment I first met my father. 'The Captain' was returning after a long period abroad, and with the exception of the first few days after my birth, this was to be the first time I would meet the man about whom my mother and siblings spoke so respectfully. I was three years old. I had been dressed in my Sunday best, and stood in the foyer of our home. Before long, the front door opened to reveal a tall figure in dark blue uniform with a long thick beard. I screamed and ran to my mother, burying my head in her skirt and all she could get out of me was '*fevvers*', my description of the beard on the strange man. Despite this, my father and I became friends soon after he shaved off the '*fevvers*', but sadly it wasn't long until he left again.

My mother, Aida Monica Cooke, always went by the name Monica. She was a bright and independent woman. Her father, my grandfather, Luther Cooke, was a medical doctor who believed education was of the utmost importance. As such he had sent my mother and her younger sister Teresa to Germany to attend one of the best boarding schools in Europe while they were still young girls. It must have been very difficult for her to leave her family and her home, and be expected to acquire a new language while at the same time perform well at school. Whether or not it was because of this, my mother was a very strong and independent woman. She would take no nonsense from anyone,

Captain James Johnson.

not even from her children and was often very strict, but in her own way she was kind and understanding.

I don't know how she coped with father away so much of the time, but I remember her sharing that when she first married, father took her home, and told her that he would be going back out to sea at the end of the fortnight. When mother asked how long he would be away for, he said he would be back in a just a few months. Mother said she had been terrified of the idea of being left in their new home alone, and noted that at that stage she couldn't even boil an egg, much less cook a meal. The night after that father left, she sat by the fire and cried so loud that before long the dog came and sat beside her and started to howl. Eventually her tears turned into laughter, of sorts, and she explained that from that moment onwards she never cried again.

Mother's sister, Teresa, visited us frequently and she and my mother would always speak to each other in German. They claimed it was to keep up their skills, but I rather think it was so that we wouldn't know what they were talking about. It never bothered me at the time, but in the years that followed, I regretted having wasted the opportunity to learn such an important language. The two sisters often spent long afternoons singing songs and playing duets on the piano, something at which they were both rather proficient.

On the days when Teresa didn't visit, mother used to sit on her own with her oils and paint for hours on end, capturing landscapes, flowers and birds. One painting I particularly remember hung in the front foyer of our house. It portrayed a river and on the far bank was a hill at the top of which was a rounded castle and an old church. Many years later when I was visiting Germany, I came across the exact scene near her school and was amazed at the uncanny likeness that she had captured, and the memory she had carried so accurately in her mind for so many years.

Mother was always constructively occupied and had produced more oil paintings than one would need to fill a gallery, and although she was at home almost all the time, she was not often directly involved with us children. Whether this was the legacy of her attending boarding school from such a young age, or her frail health or simply the lifestyle to which she had become accustomed to from her father's well staffed live in medical practice, I cannot say, but she was content enough with her own company and hobbies. She didn't go out much, and although I am not sure in what way she was ill, she never had a lot of energy and from time to time suffered abdominal pain. Fortunately she did not need to earn an income, as raising five children together with running the household staff kept her sufficiently occupied. Indeed, we were one of the fortunate families with full time staff, the most important of which was without doubt my nanny, Mary Durkin.

Mary was my world. She had been with the family for several years before I was born, but from that moment on, with mother's poor health and my older siblings at school, she spent most of her time raising me. Mary was a very warm, kind and gentle young woman in her early twenties with a beautiful Irish accent. She was slender and rather pretty with her long dark hair pinned back in a loose bun. She usually wore the ordinary style pinafore of the day with two large front pockets perfect for keeping small sweets for when I was good, which was surprisingly often. Everything about her nature was gentle. She wasn't a bit strict, and so from a young age I learnt that I could be naughty with her, but strangely, it was her kindness that made me want to behave. I didn't want to disappoint her, and it only took a slight frown to let me know if something was amiss.

Apart from Mary, those who were most influential in my young world were of course my four older siblings, although in reality I spent very little time with the two eldest due to the large age gaps. My eldest sister Mary Monica, whom we called Monica, was born in 1905, and during the early years of my childhood she was busy learning how to become a school teacher. She was always very kind to me, although she was usually busy and spent little time at home. Monica also had a lot of commitments to attend, and seemed to enjoy going out to various dances and social events in the village. When she was home, she would often sit and paint landscape scenes with my mother. My brother, John Gerard, was born in 1908 and despite the twelve-year age gap, we both got along very well. He attended boy scouts and this meant that of the few weeks of the year that he was not away at boarding school, he was often away camping. Nonetheless, when we did see each other he always tried to teach me as much as he could and a lot of what he had been taught at scouts, so he showed me how to tie various knots and even some bush survival skills, which he joked would come in useful if I ever found myself lost in Africa.

It was a quiet sort of life and we had to make our own entertainment, so I spent a lot more time playing with my two youngest sisters Elizabeth Christine, born in 1911 and Aida Frances born in 1914. Although they too were eventually sent off to boarding school, for at least the first years of my life, I was able to be with them and we spent a lot of time inventing games, making puppet shows for each other, reading stories, and playing with our pets.

Home was on the outskirts of the historic village of Ormskirk, about 13 miles north of Liverpool. My father told me that Ormskirk was named after the Vikings, but much to my disappointment I never saw anything as exciting as that! It was a town whose location featured on many of the old trade routes and there were a great number of travellers and inns. Despite the transient population, the village had not lost its traditional character and heritage. People did their best to preserve the history of the town and were very proud of who had lived there over the years, especially the nobility who had favoured it as a place for summer residencies. It was pointed out to me, on frequent occasions, that Lord Stanley lived in the mansion house on St Helens Road, right across from our home—not that the two residencies were comparable in many other ways. My thirteen year old sister, Elizabeth, taught me that the road running through the centre of the town, from Cheshire in the South, through nearby Wigan and onto Lancaster in the North was laid nearly 2000 years ago by the Romans. She went on to say that others had been fortunate enough to find Roman coins along the roadside, and that if I walked without saying a word and concentrated on looking for them, I might be lucky enough to find some, but only if I was silent.

In the pedestrian centre of the town, by the tall clock tower, was the open-air market which we visited each week. There would have been more than a hundred stalls most of which were run by local farmers and agricultural workers selling fresh produce. Mary, and occasionally my mother, would go shopping here and at the end of a trip, if I had been good, I would be given a slice of hand made fudge or some toffee sweets.

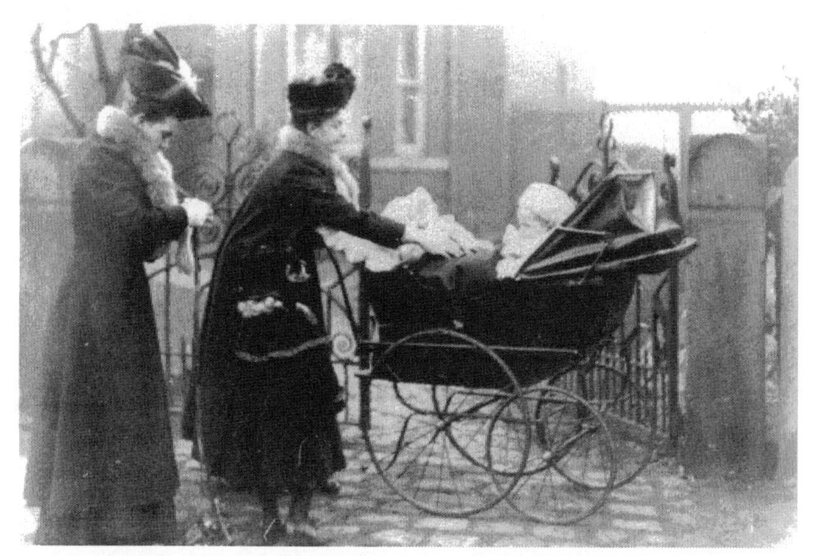

Aunt Teresa and Aida Monica Cooke, pushing my brother John in the family's pram (1908).

There were relatively few residents on the outskirts of Ormskirk, and particularly where we lived, everybody knew everybody else, or at least everybody else's business. It was the sort of village with enormous hustle and bustle during the week days, but every weekend when most of the business travellers had gone home, it became a traditional English country town, complete with the parish church bells ringing their age old carillon calling the people to worship, on Sundays.

We lived in a pleasant house named '*Eldrig*'. I wish I knew where the name came from, perhaps the Old Norse—'*Fire and King*', or simply it could have been '*girdle*' in reverse, given the family's sense of humor. Oddly enough, for all the grandeur of the double storey brick home, it was the staircase that impressed me most as a small child. The house was set out such that from the front foyer one could see across a wide open living room to the bottom of the double width stair case, made of dark and solid timber with a beautifully polished banister. Up the stairs, behind the railings were all the bedrooms, each overlooking the entrance foyer and grand stair case. Half way up there was a landing, before the stairs divided to the left and the right, with the younger children's and nanny's bedrooms being off to the left and the parent's rooms off to the right.

The sun was always shining into our house and garden. It was a wonderful property, which probably was only of average size for the time, but seemed to me to be a vast and never ending wonderland. Along the front of our garden, which bordered the road to Liverpool, were tall trees that reached to the heavens. I was too small to even think of climbing them and spent most of my time in the garden viewing the world from my pram. I did all that I could to leave the confines of the pram. I was ready to explore. The pram was large with a well in the middle, the lid to which could be removed to provide a seat at each end. I know this because I discovered that I could remove the lid to reveal the bottom of the pram which was covered in small black and white checks, but sadly, that was as far down as I could burrow, there was no escape hatch below. For now I had to be content to gaze up at the sky-high trees.

Mary Durkin had been a part of my life for as long as I could remember and although she worked mostly as my nanny, she also assisted in the kitchen. When she wasn't busy doing chores, she would take me for walks through intriguing places with little paths, which meandered up and down through the woods or gardens. Sometimes she would take me to the park, which was directly opposite our house. The park was surrounded by tall wrought iron railings and magnificent large gates, which were directly opposite our front door. (All of this was removed in 1939 to contribute to the war effort). The paths in the park were different to the little tracks in the woods. Here they were wide gravel roads, well defined, with an air of upper class superiority, not that that meant a lot to such a young boy. Regardless of which park we chose to walk in, the tall trees and long shadows meant it was cold most months of the year.

Mary would dress me in little leggings fastened by a long row of tiny buttons, which required the use of a button-hook (which I still have). Life was a much slower pace back then, and despite the lack of gizmos and gadgets to make life easier, people tended to have more time to bother with such fastidious details.

One occasion left an indelible mark on my 4-year-old mind. On this day, Mary was a little late with our walk and when we returned to the gate we found it closed and locked. Whether Mary was able to attract attention, or that someone in the family noticed our absence I cannot say, but we were rescued by means of a tall step ladder; an ignominious exit for Mary but a day of excitement for me.

At this age, my life was surrounded by fences and gates, through which I could always see just enough to know that I wanted more. At home, our front gate was made of wooden palings with a rounded top. The bottom rail was such that even a small boy could stand on it and peer out. I spent countless hours watching and wondering about the world out there. It was from here that I had my first glimpse of royalty when the Prince of Wales, who would later become King Edward VIII, drove past. Seeing me at the gate he waved to me and I to him. I was a nobody to him of course, but then, I didn't know who he was either, so we had that much in common.

Another vantage point was the bedroom window, which overlooked the front gate and gave full views of the road. It was here that I saw a cousin, either Eric or Gerard Fishwick, who had called to see us. He had travelled on a motorbike, and was now about to leave. He sat idling on the footpath, waving goodbye to my mother and then accelerated with the speed expected of a young lad. Just as quickly, both he and the bike came to a screaming halt. He wobbled a lot, and I am sure he would have fallen, had it not been for the stretched jacket that now pinned him so firmly to the bike. His raincoat had gotten caught in the valves of the engine. My mother had to dash off and find a pair of shears to cut him loose, while he tried to maintain his awkward position on the bike. All this enabled me ample time to observe the little engine. It was the first time I had ever noticed an engine or that it was made of movable parts. It was fascinating.

Frances, the youngest of my three sisters, and I spent a great deal of time watching the outside world from the sash window, where we could see the intersection of two main roads and a nearby bus stop. In those days there was very little traffic. We would see the odd horse and cart or modern motor car, but more commonly we would observe cyclists or men in suits with official looking briefcases dashing to catch the bus, and ladies who always seemed to be carrying something.

One afternoon, on the road running along the front of the house, there was a car coming from the direction of Liverpool. On the other road, St Helens, which ran along the side of the house, a motorbike was approaching. 'Look', said my sister, 'watch these two have an accident'. Just as she had predicted they did, right before our front gate. Suddenly, the motor-cyclist was lying on the footpath propped up against the big red letter box which stood near the corner. He clutched his foot and hollowed in pain exclaiming words I had never heard before. It was the most exciting day we had had in a long time.

From the many hours spent watching the road through this window I learnt quite a lot about motor bikes. Strangely I was never particularly interested in riding one, but as a four-year-old boy, I became increasingly fascinated at how they worked. I wondered what drove them and what made the pistons move up and down and why they had so much power. I asked Mary and mother about these things, but neither could sufficiently enlighten me, so I decided to wait until my brother John was home from boarding school, but before then, an opportunity arose for me to learn more.

My mother's father, Doctor Luther Cooke, came to visit us and as always, he came in his motorcar—'the Buick'. With father away so much and mother not able to drive, there were very few occasions on which I could ride in a motorcar, but I had often admired grandfather's car, and on a few occasions had even travelled in it.

On this visit, my mother told my grandfather that I had been asking about engines. Grandfather was delighted and took the time to show me his car and how it worked, and he tried to explain what made it go. I was so thrilled.

The 1910 Buick was a Laudaulette model 'open tourer', which meant it had the chassis of a limousine with a convertible soft top. It was the type that was used in public processions and as grandfather would say, it offered 'alfresco motoring at its finest'. Only a short time before my baptism, it had been purchased to replace the doctor's horse and carriage and now it was his pride and joy. Grandfather wasn't particularly interested in being chauffeur driven, but instead he loved to sit up front, in the separate chauffeurs section, where he could experience the thrill of driving a motorcar. The only problem was that grandfather, being such a loyal man, could not bring himself to dismiss his driver, Angus Ball. Instead, he employed him full time to look after the retired horses and grandfather often drove his own driver around. For me this worked out very well because Angus had a son, Francis, who was my age and equally energetic and curious about motor cars and the fascinating world of engines.

Cyril's Grandfather's 1910 model Buick EK-78, on one of the rare occasions that it was driven by the chauffeur, Angus Ball.

Chapter Two
More Questions than Answers

I WAS FIVE YEARS OLD when my world changed dramatically. All of our furniture was devoured by some huge wheeled metal monster which Mary called a removalist truck. Mary and I took the train with the rest of the family to a new life. We were to settle on the property of my maternal grandfather, Doctor Luther Cooke, in Aspull, about 12 miles east of Ormskirk. It was an enormous property on which he had two very large houses, two tennis courts and a few horses, but promises of this did not comfort me. When we arrived, to our new home, the

View across the Aspull moorlands adjacent to Cyril's grandfather's estate.

world looked cold, wet and windy and not a bit interesting. The inside of the house was a muddle of furniture and looked no better. But Mary, when she could spare a moment from her other duties, would dry my eyes and assure me that the world had not come to an end.

The general strike of the mid 1920s had brought my father's seafaring life to an end. It was at this time that many ships were laid up or scrapped. My mother's illness and a growing family made even bigger demands on my father's resources. So my grandfather, the local doctor, helped out by employing my father to assist in the practice, particularly in his pharmacy dispensary business and by providing a large house for us to live in. My father's father had been a chemist, and not that I suppose that this was an automatically inherited skill, but it seemed that my father did in fact have the discipline and precision needed to run the dispensary. I suspect it was intended to be only a temporary situation but it lasted for the rest of my father's life.

In many ways, Aspull was very similar to the neighboring but larger town of Wigan, as it was only a short distance away, and shared a similar history. They were both small communities in which everybody knew everybody else, with the notable difference being that most people in Aspull were miners and with only two primary employers, many of them also worked together. There were also far fewer prosperous people in Aspull, not that that bothered me in the slightest, but it did mean less well manicured parks in which to go walking, and this did disappoint me. However, Aspull was a beautiful township, on the edge of Lord Crawford's estate and the town itself was spread over a large area because much of it was on moorlands. In exchange for the wide gravel paths that I was used to in the park, I was delighted to learn that the vast expanse of moorland was all Crown Land, and open to everyone, from agisters to hikers. This meant that there were endless fields with open access and one could go walking for miles through interesting countryside, making their own trails, when it wasn't too boggy. It was a good thing that I was now big enough not to need a pram, and could do it all by myself, so long as Mary came with me.

The town had a population of 10,000 and was settled in

the days of the horse and cart, as was evidenced by the wide meandering roads and tie-ups in front of the pubs and shops. Perhaps more interestingly, at least for a young boy who was fascinated by engines, was that the railway track running through the town via the coal mines and onto Liverpool, was built long before the first steam train was even invented! A distant relative, whose father had been a surveyor for George Stevenson, explained to me that a long time ago, it had been understood that a cart brim full with heavy coal would take fewer horses to pull if it were on a track. Thus they built the railway line quite some time before the steam engines came to relieve the horses, but by the time my grandfather's family had settled in Aspull, in the mid 1800s, Stevenson's locomotive invention had been wholeheartedly embraced as a means of transporting minerals and passengers along the tracks—all powered by steam. I was so intrigued that I used to ask Mary to take me for a walk to the railway station, just to watch the trains go by.

I enjoyed settling into the new village, and noticed that mother seemed quite happy about returning to her family's property. It was a magnificent establishment, and in a sense it was its own community, with about 25 people living on the property. Doctor Cooke and his wife lived in a very large home closest to the entrance of the property. Inside this house, the front rooms and the main living room were all dedicated to his surgery, forming the waiting room and the consulting rooms. In addition, it was from here that he operated his dispensary and chemist shop.

Aside this first house was a steep embankment, which ran down onto the flat manicured lawns, that formed grandfather's grass tennis court. Behind this was a rose garden, which was properly cared for, and produced many brightly coloured petals. On the other side of the roses was a newly surfaced tennis court, which was boarded on all four sides by a raised curb. I remember this tennis court particularly well because a few winters after we first arrived, my friend Francis Ball and I decided that it would be a good idea to block the gutters and run the garden hose, such that we could flood the tennis court and the overnight freezing temperatures would form an ice skating rink. This we succeeded in doing, but not without totally destroying the surface of the tennis court—an incident for which I was severely reprimanded. It wasn't until many years later that I could enjoy a game of tennis without the guilt, and even then I chose to play on the grass courts.

Doctor William E Cooke working in Luther Cooke's chemist dispensary.

Beyond the tennis courts were two glass houses in which grandfather would grow a variety of vegetables and some wine grapes. These ran close to the side of a double storey garage large enough to house five or six cars and the old cart, which was no longer used, as well as some additional storage amongst the rafters. Directly opposite this garage was a paddock in which

Luther Cooke hosted the church garden picnic each year in his field alongside his house and surgery.
(Note the number of chimneys on the rooftops).

The lawn in front of 'Ashfield House', Luther Cooke's home and practice, showing the rooms dedicated to the surgical practice and chemist shop.

the retired horses now stood, and from time to time cows would appear in there, although I am not sure whose they were. In addition to this was a small hobby farm pen, in which we had a great many chickens, and the odd seasonal turkey. It was in this section of the property, removed from the business end at the front of the property, that there was a very small cottage, in which an elderly woman lived. She had been my mother's nanny, and in her aging years, the family looked after her. Every day, someone was sent to check on her, bring meals, light the fire and spend a little time with her.

Aside the tennis courts, and back up towards the front of the property, were three cottages, two of which were adjoining. In the first small adjoining cottage, lived the Durkin family, from whom Mary had descended. The Durkins were a large extended family, and everyone seemed to be a 'cousin'. They were all employed, as cooks, cleaners, gardeners and maids and surgery staff, on the property. In the other cottage lived the Ball family. Angus Ball had been grandfather's coachman, and was still employed as his driver. Angus and his wife had a son, Francis

Luther Cooke with his chauffeur, Angus Ball, standing in front of the medical practice 'Ashfield House'. (Date unknown, c.1910).

Cyril and cousin Peter walking through the lush fruit and vegetable garden.

Elizabeth Christine Johnson tending to her bee hives.

Ball, who was my age and we soon formed a life long friendship. In the third stand alone cottage which was originally built for the assistant doctor, lived my mother's sister, Teresa and her three children.

Near the cottages there was also a very large fruit orchard and vegetable garden, in which we grew nearly all the fruit and vegetables and herbs needed for everyone living on the property. Despite the cold winters, the orchard produced an abundance of everything, and no doubt my sister Elizabeth's bee keeping hobby helped in its success, indeed her bees were busy enough to produced over 90 pounds of honey in one year. Although we had gardeners, we often pitched in and helped with the garden, which was both a chore and a hobby.

In addition to all this, behind the cottages, was a very large house similar in size to that of the doctor's. It was here that my parents, brothers, three sisters and I lived. It was a very busy and active property.

Before we had arrived, grandfather had ordered the installation of a private telephone line. Between each house and cottage was a telephone line. This was not something that I was permitted to use, but I watched mother use it countless times. It was one of the very early telephones, with an earpiece that was held up to one ear and a separate microphone out in front. To make a call one had to wind the handle on a small box, which worked like a generator, to ring the bell at the other end of the line. This worked wonderfully and was good for sending quick messages to people on the property, but it was not until much later that we connected with the newer telephone lines, which operated via an exchange and connected to anyone in the village.

Mother and her family were well known in the village, so when we attended church on our first Sunday in Aspull, we were welcomed as though we were long lost friends. Although the preacher seemed more excited to meet my siblings and me than any of us were to meet him. We lined up, dressed in our Sunday bests, and shook hands with the black and white man, and I immediately looked around to see if there were any boys my own age.

Those that were, were being escorted closely by their own parents, so there was little fun to be had there and one look inside told me it was obviously a very formal and traditional church.

Previous page and above: Two photographs of
Our Lady of the Immaculate Conception Church, Haig Road, Aspull.
(Roman Catholic—in case the name didn't give it away!)

I was now five years old, but try as I may, it was difficult to sit through church quietly. It was said that I was rather cute, with white blonde hair and blue eyes, and that this somehow covered a multitude of sins. But despite this, I remember two incidents that disgraced my family. Firstly, I had found some of father's old West African coin collection, and decided to place a coin from that into the offering bowl as it was passed around one Sunday. The next week the preacher announced the amount that was collected in the offering, including the extra 'one tenth of a penny' in the total. He looked over his pulpit and stared directly at me. My brother John got it in an instant, and his eyes lit up with the pride of an encouraging older brother. John turned to me and we both sniggered. I tried to cover my broad and delighted five-year-old grin with both hands, but one glimpse at my mother's blushing face made me realise that this was no disguise. Fortunately there emerged a few chuckles from the congregation, and my mother's posture quickly relaxed and I even saw her smile—when she thought we weren't looking.

There was little follow up on that incident, which paled into insignificance compared to my later performance. Mother's family friends had invited us to their wedding. We all attended, except for father and Mary. Mother, having her hands full with five children, chose a row near the back. The same preacher was conducting the ceremony, and had said something that didn't make sense to me. Seeing the bride crying I was concerned. When the preacher asked if the man would take this woman to become an 'awful wedded wife' something sounded horribly wrong. I was so indignant that I climbed up onto my pew, stood, and boldly shouted, 'No'. There were a few gasps, followed by a moment of shocked silence, until one or two chuckles burst forth into roaring laughter across the whole congregation. The entire congregation turned around and stared at my family and me.

A firm tug on the arm got me quickly off the pew and one look at mother's face made me realise that this was much worse than anything I had ever done before. I felt heat rushing into my cheeks and glanced up at the preacher, who did look a bit surprised, but quickly resumed as though nothing had happened. We left immediately after the ceremony and by the time we got home, the incident was still fresh in mother's mind. I got a very strong lecture about not butting into other people's business and not expressing an opinion on things that I did not understand, but worse than that, my two sisters Elizabeth and Frances brought it up time and time again. It took years to live it down and from then on, I was often accused, rightly or wrongly, of saying things out of turn.

I think Mary must have heard about the incident, because the next day she avoided the subject of the wedding altogether. In any event, she had her hands full. There was always a great deal of cleaning and laundry to do, and in addition Mary seemed to be constantly baking for an endless procession of friends and neighbours who came to visit.

Our kitchen was large and always warm, so I spent a lot of time there during that first winter. Along one wall extended a wooden cupboard from the top of the very high ceiling right down to the floorboards. At the end of the row, nearest the kitchen fire, was a cupboard where I kept some toys and could play under the watchful eye of Mary. All around the kitchen, the cupboards had sliding doors, perfect for a little boy to shut himself in, and host many imaginative adventures. The wooden floor of the bottom cupboard was about six inches above the level of the kitchen floor and the boards in the cupboard did not quite reach the back wall, leaving a small gap down which I was able to post pieces of paper, playing cards or anything else which was thin enough.

The kitchen fire was always burning. Although it may have rivaled the piano room as the heart of the home, it was the place where Mary and I spent most of our time. The fire was a stand alone, all in one hot box, and everything was made of wrought iron. In its belly was a place for burning coal, with a vent pipe just above. Immediately in front of the vent, and directly above where the fire burnt, was a cooking surface with five or six hot plates and on the edge of this were a series of little rails from which one could hang towels. Below the smaller hotplates, directly aside the fire was a double door oven, large enough for a Christmas turkey or three loaves of bread. Below the fire and the oven were other doors which concealed a damper control, a soot clean out and an additional warming place, which if left open, would be half occupied by the cat.

The fire, although effective for cooking, was not enough to keep more than one room of the house warm during the cold English winters. Fortunately we had a fireplace in every room of the house and along the rooftop could be seen a total of seventeen chimneys. It was a good thing we lived in a coal-mining village, and little wonder that the man who delivered coal had to come every week. The coalman would come to the back door, covered in black, and haul a large sac out from his horse drawn cart and then drag a second sack all the way to the kitchen door. Near where he would stop his horse, lived our African parrot, father's bird who talked incessantly and was very good at learning new words. The parrot was always learning something new as evidenced by its repertoire of swear words thanks to some of the sailors who had been on father's ship. One day, after watching the routine so many times and with the coalman off near the kitchen, the parrot said 'giddy-up' to the horse. The horse immediately took off down the property with the coalman running after it. It was very exciting and I often found that animals provided some of the best entertainment.

When Mary was busy attending to chores, or my siblings were not available to play, I would turn my attentions to 'Bunny', who was in fact, a very friendly cat. She was black and white with three socks. To most people, Bunny was just a common cat, but to me she was my confidante and playmate. We spent countless hours together by the kitchen fire. Bunny was a gentle cat and would let me tow her around in a little wooden box attached to a piece of string, and she sat perfectly still. At nights she would sleep in my bed and keep me warm, and I would bury my small fingers in her fur. She never scratched me despite all the games I used to play with her.

Like me, Bunny was a born explorer. She too liked to watch and wonder at the outside world, but seemed to have a much easier time than I did getting up on top of the fence posts and climbing trees. One day when I was busy inside, Bunny went outside and sat on the garden fence where she could see the neighboring cottage. Later that afternoon, I went looking for her. I searched inside the house, and then walked through our large garden, calling her name. It was unusual for her not to appear when called, so Mary let me put a little milk in a saucer, but I never saw my friend again. The next morning, my mother offered a partial explanation. That was my first encounter with death, the cause of many tears for a five-year-old boy. Years later I was told that a bad tempered neighbour had thrown a stone at her and leaping to get away, Bunny had fallen onto some barbed wire.

I was sad for many days until one morning, my oldest sister Monica came and woke me up by slipping a little kitten into my bed. That helped a lot. She had always been very good with animals and children, and seemed to know how to cheer me up.

This cat was mostly white, except for a black patch on its two back legs and a small little patch of black under its nose that looked like a moustache. It was very cute. In the days that followed, while trying to decide on a name, someone in the family suggested we name him after a German politician who we had frequently seen in the newspapers. A man with a moustache. I agreed and so the name was settled upon. This was fine in 1925, but not such a good choice by 1939 when mother had to stand on the front porch calling for Adolph to come home.

It was in these early years that I was also given two bantam hens, Phoebe and Phyllis. They laid lovely little eggs, which regularly contributed to my diet. These were not like the chickens in the main hutch, they were much smaller, and I could easily take them out of their cage and carry them around and put them on a fresh piece of lawn. They were very friendly and I was sure they understood what I said to them. Sadly however, one day they stopped laying—having a rest I suppose. Unbeknown to me they too were consigned to be part of my diet. What made matters worse was that in the process of being prepared for the oven it was discovered that they were full of eggs. A sad anatomy lesson. I had lost my friends, and no one seemed to want to talk about it.

Following this, for whatever reason, someone gave me a

plump, pure white Angora rabbit. I soon had seven rabbits, and was entirely unaware of where they had come from. This was not for a lack of questions on my behalf. There was an attempt at some explanation, but it didn't register much with anything I had ever heard of before and the more I asked questions, the more vague the responses became. It seemed as though there were some subjects that were only for grownups and I don't think my ten year old sister, Frances, even knew much about them.

One day, Mary took me to the 1925 annual Wigan Country Fair. I can recall it all so vividly. There were stalls selling all sorts of brightly coloured things, and there were swings, merry-go-rounds, and entertainers. The food was fantastic, with many new flavours and varieties of sweets to try. Most impressive of all, there was a display of large traction engines with smoking chimneys and belts driving generators, which provided power and lights to the fairground.

I recalled the explanation of the steam train, and the parts of the motor-bike engine which had gone up and down and caught my cousin's coat and I wondered how all these engines worked. It was incredible to just stand and watch, and trace with my eyes where each part led to, and try to figure out what made all the pieces move. It was fascinating.

I came home so excited and wanted to tell everybody about what I had seen, and the sample bags I had collected. Little did I know that I had brought home something extra that day. Within the week I had developed a burning throat and ruby rash that was easily identified as scarlet fever. But that was not all.

I awoke one night soon after, screaming with pain and was unable to move my legs. My grandfather, Doctor Luther Cooke, examined me and then ushered my mother into the hallway where he informed her that this was rheumatic fever—an inflammatory condition which had developed subsequent to the scarlet fever, and was likely to affect my heart. I was not told much about any of this, but I knew my mother was worried. Just as had been decided with the cat, the chickens and the rabbit, this too was not a topic for discussing in detail with a five year old boy.

There was clearly something going on, and it was sufficient

to alarm my mother, but very few of my questions were answered. Strangely enough, I think it was the 'not knowing' that made it all the worse.

The treatment was salicylates, with its nauseatingly sweet taste, and bed rest, and so I was banished to my room for what seemed an eternity. Not that I knew it at the time but I was one of the lucky ones, as many other children were sent off to special foster homes for anywhere from several weeks to a few months of bed rest with this frightful illness. Fortunately, in my room, I was also allowed to have my toys and my favorite picture books, which now meant everything in the world to me. A few weeks later it was decided that I could have the family Pathe gramophone in my room. This cheered me up greatly and made me feel very special, as it was said that my father had been given it as a gift from one of his sea passengers (although I now really think he won it playing cards, something at which he was an expert). I almost wore out two classical music records, and another of peculiar dance music, which father must have acquired in Africa. At the end of this record, over the applauding, an officious sounding English voice boomed, 'Thank you, white men this way, Niggers that way'. Perhaps because it was the only talking I heard for most of my days, I found this hilarious and practiced imitating the man's serious voice, and even developed arm actions for it. When grandfather came to check up on me one day, I was promptly educated about the matter and told never to repeat the phrase.

When I was pronounced 'cured', my room was sealed with a white tape and its contents (those which could not be washed or burned) were fumigated. Sadly, it was decided that my paper books and all my soft toys, my most precious worldly goods, could not be suitably fumigated, and so some obliging family member went and set the lot on fire! It was devastating. I was grateful the cat had already died! Yet it was not only my toys that I had lost, but something of myself, for I later found that my heart was affected and I would go on to develop a leaky valve. Yet worse than all this, I had lost my Mary. Mary had gone away and whether this was because she had been thought responsible for exposing me to the infection or because of the reduction in the family income I never knew. I was devastated. Mary had meant everything to me, and hearing that she would not be coming back, felt as though I had lost my own mother. My heart ached worse than anything ever had during the sickness.

I later found out that during the 1920s rheumatic fever had been the leading cause of death in children my age. I wondered if any of the other children had known anything more about this illness during their sufferings, although I suspected that they too were probably 'protected' from the truth, and left anxious and frightened with unanswered questions. It seemed as though the divide between children and adults was universal, and there was an ever-growing range of subjects about which we were not to be informed. This was confirmed when I finally went to school and nobody there could enlighten me on my brewing questions.

Growing up I continued to have more than my share of illness and accidents. One wet and windy day, a few months after the scarlet fever, I was allowed to bring my scooter indoors. There was plenty of room for me to ride round the large kitchen table. As my confidence increased so did my speed until the inevitable happened: I crashed. I connected with the wooden cross bar that was the handle and steering column. I landed face first and broke my nose. At first I saw a few spots of white, which were quickly replaced by tears of pain, and then tears of self-pity. Mother tried to comfort me, but it only made me miss Mary more. I was angry with myself for the crash, and I began to worry that it was my fault that Mary had been sent away.

On top of everything else, I soon developed tonsillitis. This, in light of my recent ill health, was considered to be a very serious development. With a burning throat that reminded me of the scarlet fever, my biggest concern was that I would be banished to my room again, a terrible thought, given how few toys I now had! I curled into my mother, with my head on her lap, and she sat with me by the kitchen fire. I was now six years old and I think my family had given me up, for I was told not to be afraid and that heaven was a good place. All of which did little to encourage me.

Fortunately, most of our family seemed to be medical doctors, the most brilliant of whom was my uncle, Doctor

Cyril with
Doctor William E Cooke,
'Uncle Billy'.

William Cooke, the son of Doctor Luther Cooke. It was this 'Uncle Billy' who came to inspect my latest development and he considered that instant action was called for. The kitchen table was promptly scrubbed up for the job. Chloroform was poured onto some cotton wool and this was then placed over my face. For a moment I felt I was drowning in a sea of cotton wool or was this supposed to be heaven? My tonsils were removed, and I survived, but I think it was a surprise to everyone, and an absolute credit to the doctor.

When I say Doctor William E. Cooke was brilliant, he really was. Apart from being the Director of the Pathological Department at Wigan Infirmatory, he was also a Fellow of the Royal College of Physicians, London. In total, five books—mostly on microbiology and hematology—and thirty publications in major medical journals, stand to his credit. In 1924, the *British Medical Journal* published an article of his, entitled 'Fibrosis Of The Lungs Due To The Inhalation Of Asbestos Dust'. Doctor Cooke was the first to prove and publish this causal link that others had long suspected. Following this, in 1927, he went on to write another article entitled 'Pulmonary Asbestosis' which was also published in the British Medical Journal. As a result of Cooke's article, and his testimony at the inquest into the death of an asbestos worker, the British parliament commissioned an inquiry into the effects of asbestos particles. The report—entitled 'Occurrence of Pulmonary Fibrosis & Other Pulmonary Affections in Asbestos Workers' was presented in March 1930. It concluded that the asbestosis was irrefutably linked to prolonged inhalation of asbestos dust. In this report the first cohort study done on the health of asbestos workers was presented. It found that 66% of those employed for 20 years or more suffered from asbestosis. The report led to the publication of the first British Asbestos Industry Regulations, coming into effect in March 1932.

Chapter Three
Elementary School

UNDER THE ORDERS OF 'UNCLE Billy', I was not to go back to school for a long time after the tonsillitis. My convalescence seemed to be endless, but not unpleasant. When my two younger sisters were on holiday from their boarding school they played with me. Frances had a vivid imagination like myself. To us, a Ludo board became a school with four classrooms joined by corridors. Little people were, of course, the coloured tokens from this game and the Halma game, and we used a matchbox for a motorcar. I enjoyed my time at home, but when the girls returned to school at the end of the holidays, I did everything I could to convince my parents, and Uncle Billy, that I was ready for school. I had no idea what I was bringing upon myself.

I was sent to the village elementary parish school, Our Lady of Grace. On this first and momentous day my mother escorted me there and left me in the hands of a Miss Giblin. It became apparent to all that I was not as brave as I had thought. I remember yelling as if I was being murdered but there was no escape. The gates were closed. Miss Giblin took no nonsense, and I was shown where I was to hang my hat and coat, some way down a long narrow cloakroom. Adding to the unfriendly appearance, the walls were painted in dark green enamel and on one side was a row of bulbous hat pegs.

The school comprised of two large rooms; the first room contained three junior classes and the second room contained two older classes, with 13 years being the upper age limit. The classrooms were mostly made of white plastered walls, with a continuation of the dark green enamel paint around every door and window frame, and long blackboards that covered most of the front of the room, but at least these rooms had the redeeming features of brightly coloured birds and flowers painted above the window frames.

There were only two qualified teachers in the whole school, Mrs. Ireland and Miss Bridget Giblin who worked in the first room. In the second room, two ladies supervised the older children, with Gertrude Moore also being the head mistress.

I remember the first lesson well. It consisted of twenty-six large squares of brown paper, on each of which was a letter of the alphabet. The students in my class sat in a circle and were each given some dried corn. We then took it in turns to place the corn on whatever letter the teacher called out. This I presume was to teach us the alphabet, or at least recognise what shape the letters were, but I was in a dark mood, and did not do the family proud. For all my pouting, it did not take me long to realise that there was no use in protesting, the arrangement was firm and would last a long time, many years in fact, judging by the age of the children in the room next door.

Once I decided to apply myself, I was rewarded with great encouragement from my teachers, who must have thought I wasn't too bad, because by the end of my first year I had progressed onto the 'other' room with the older children. Here I actually enjoyed school. We were read stories of ancient Greece and Troy. The adventures of the gods and heroes fired my imagination, and I wondered where one could get a pair of Hermes flying shoes and thought how amazing it must be to be able to fly.

On Friday afternoons, if we had been good, which of course we always were, we were given two or three small sticky sweets. These would be screwed up in the corner of a dirty handkerchief to be eaten at a later time, usually on the way home. I was trusted to walk home with my friend Francis Ball—a distance of a little less than a half a mile. That distance seemed much longer to two little boys. En route there was a ditch to be explored and we were often scolded for being late home for dinner.

It was at school where I first met girls (excluding my three sisters who didn't count of course). There were two in particular that I remembered for the rest of my life; Kitty Moore and Lilly

Moss. Kitty was the daughter of a local haulage contractor. She was beautiful, with long curly mousey brown hair, and she often wore a knitted cardigan made from many different soft colours. Kitty was always very kind. At about age seven I fell in love with Kitty, but alas my love was unrequited. That year, the school put on a Christmas performance to which the whole village was invited. Kitty and Lilly sang a song, which was about a quarrel between two girls, 'You shan't play in our yard, you shan't climb our apple tree'. In years to come Lilly remembered this and sang it for me on a cassette tape. It was a very sad day when I learned that Kitty had pneumonia and had died: she was only seven years old.

Lilly Moss selected as the May Queen, 1927.

Lilly Moss and I became very close friends as the years went on, although as she would later confess, at that age she was terribly shy whenever I was around. To me she was the most beautiful girl I had ever known, so it was no surprise when she was chosen to be the May Queen in 1927. Her role was to lead the school procession in celebration of the beginning of summer, and as was the tradition, she was dressed in white and crowned with flowers. I was so excited for her that I even learnt and memorised Lord Tennyson's poem, The May Queen,

> *"You must wake and call me early, call me early, mother dear;*
> *Tomorrow 'll be the happiest time of all the glad new-year,*
> *Of all the glad new-year, mother, the maddest, the merriest day;*
> *For I'm to be the Queen o' the May, mother, I'm to be the Queen o' the May".*

In our village most of the children came from poor families. The parents were mainly coal miners and the strike of 1924/5, although now over, had left them in poor circumstances. The workers wore clogs, which were very serviceable, and the replacement of clog irons was relatively cheap. Most children wore clogs also, even some of those from the few financially better off families. Clogs had thick wooden soles and were warm in winter even if they were not beautiful. The irons acted as skates on ice and required careful control. The village clog maker was Steve Balfour and he could replace clog irons while you waited, if he had not gone off on a drinking spree—which happened often. In order to be in the fashion, I was given a pair of clogs but I suspect that the sight of the Doctor's grandson wearing clogs was regarded by some of the family as out of character, and my clogs did not last long.

The ladies of the house in their 1920's outfits.

Fashion in those days, although changing rapidly, was a clear communicator of class, and there were just some rules that could not be broken. Of course, as a young boy, it didn't bother me one way or another and my wardrobe was always very simple. On casual days, messing about the house or garden, I would wear a shirt and short trousers, with a knitted red or black pullover. For special occasions, such as attending church, I would be dressed in a proper suit, usually my grey one, with a jacket and tie and of course, polished shoes. Like most boys then, I always wore short trousers until I was twelve, and from then onwards, it was always long pants. Whether the transition marked having the maturity to avoid dirty puddles, or the need to cover hairy legs, I am not sure, but that is the way it was done both in society and in all the posh schools.

For the women however, fashion was an entirely different matter. My mother now wore short dresses, above the mid point of her calf, and they often had fancy trimmings and low waist lines, some even had no waist lines, and more or less just hung. From time to time mother wore a big fur scarf and when we were going somewhere proper, she often wore a fancy hat with feathers on it or a cloche bell shaped hat made from felt. Mother smoked occasionally, and when she did she used a full length cigarette holder, which made her look very posh like the women I had seen on advertisements. Don't ask me anything about her shoes, I wouldn't know, but I do know she always had a handbag, and in the handbag was a little packet of round chocolate buttons for when we were well behaved. My sisters were also up to date with all the fashions.

Other than the fashions, children in those days had much more freedom than those of today. No one ever thought that there could be any risk to us, other than from the usual accidents so we were permitted to walk reasonable distances from home without our parents. I was sent on my first errand at age seven and was instructed to go across the moor, about a quarter of a mile, to the cottage of two elderly sisters to buy a cake of coal-tar soap. I handed over my pennies and was duly presented with the soap by a very old lady who spoke with an accent and a vocabulary of ancient Lancashire words I had never heard before. I understood almost nothing of what she said.

I must have learned something at school for on another occasion, when I was a little older, I was sent to buy something at Mrs. Eckersley's shop. On the way home I counted the change I had been given and found it was too much. I turned round and went back to tell Mrs. Eckersley of her error. She must have been impressed for she gave me a three-penny piece for 'being an honest boy'. That was a lot of money in those days, enough to buy three packets of sweets, which I promptly did. Mother was not as delighted when I arrived home with half eaten bags of sweets and I was told, they were to be shared and eaten slowly, with no more before dinner.

There was not complete freedom for a child of course and I was forbidden to wander beyond certain defined limits. One day, just beyond these limits, a boy called Clifford Monks was being given a ride on his brother's bicycle handlebar. A passing motorcar, trying to avoid the wobbly pair, then collided with the railway bridge. The front of the vehicle was dangling over the rail line, a most spectacular sight! The temptation was too great and I had to have a closer look. Unfortunately, I was spotted and spoken to most severely by my father. After that I was not allowed out of the house for two days. Being spoken to by my father made the most hardened character tremble. Although he had retired from sea, he was still by nature a true Captain, who did not take well to disobedience.

Mid way through 1927, I progressed, if that is the word, to another school, Wigan Notre Dame Convent. This was a very civilised establishment run by 'the Sisters' but which also employed lay teachers—one of whom was my Aunt Teresa. The school was highly polished, in every sense of the word, and even the floor was shiny as tends to be the way in convents. The children did not wear clogs at this school. We had indoor and outdoor shoes, which we had to change in the cloakroom several times a day with all our activities. This school was for girls but a few boys were permitted, though only in the lower forms. The Sisters were actually very nice, and so we tended to be respectful towards them. The lessons were not too taxing and I learnt very little, other than how to slide up and down the shiny corridors in my socks.

I made two very close friends at school, Bobby and Sheila. Bobby's father was Doctor Ferris, who often came to fetch Bobby in his motorcar, a Model A Ford. There were not many cars in the village in those days, so it was always exciting to look at it, and occasionally even be shown under the bonnet. It was a great adventure to go to play with Bobby at his home. It was a big house about which I mostly remember the garage, which was heaven for seven year old boys. Like my grandfather's garage it had a large upper loft as well. Up the top was a special building set for children, made of strong cardboard, sheets and boxes, each about a cubic foot. Together with the sheets and cubes, were some interlocking poles, and all of the parts had pre-cut holes in which we could easily fasten special bolts. This enabled us to build castles or slides or carts or whatever the imagination prompted. There were also ropes hanging from the garage roof, which could be used as swings, although presumably they were meant for some other purpose, as they were often covered in grease.

Teresa Moore (nee Cooke) (c.1918).

Sheila Ainsworth started to play with Bobby and me at school, and as the year went on, she used to come over and build with us, in Bobby's garage. She was a beautiful girl, and not that either of us would ever have admitted it, but I think we both fancied her. I don't think she noticed us in that way, and anyway, I still had my heart set on Lilly, not that she would consider me either. Fortunately, Sheila was always obliging when I needed help with my schoolwork, which seemed to be very often. It wasn't that homework was particularly hard, but it always interrupted my play and so it never got the time and attention it deserved. It was usually at the last minute that I would realise this, and Sheila would come to my aid. I really didn't want any of the teachers, especially Aunt Teresa, knowing that I had not done my homework properly, or word would surely make its way back to my father.

Aunt Teresa lived on the property, and so she took me to school on the bus every day. She had married Grandfather's assistant doctor some time before World War I and moved into

'Grandfather', Luther Cooke, was always very involved with his grandchildren and can be seen playing with Cyril's cousin, Peter Cooke.

a cottage with him. Doctor Moore, who rejoiced in the nickname Troggles, came from a distinguished family in Northern Ireland and numbered amongst his relations Sir Thomas Moore and a Lord Craigavon, at one time the Premier of Northern Ireland. He was well established, and had everything going for him, including a beautiful wife Teresa, and their three young children. When the First World War started Troggles had joined the Royal Army Medical Core. He blotted his copybook when he failed to return—having met a nurse and deciding that after the war he would live with her in Lorenzo Marques, a Portuguese settled part of East Africa. The disgrace of a divorce left Teresa having to work for a living and that was how she became a teacher at one of the Notre

'Grandmother' Aida Cooke, who also spent a lot of time with her grandchildren.

Dame Convents, where her other sister, Aunt Mary, had been the headmistress. Even with her teaching job, Grandfather helped their family a lot, as there were very few other supports for a single woman raising three children, especially a woman who had not been widowed, but divorced.

Teresa was an expert in the early English languages, such as Saxon and the writings of the Venerable Bede and the Doomsday Book and she was also a poet. She had attended the same German school as my mother, which explained her brilliant command of both languages. Mother and Teresa continued to speak to one another in fluent German, as a means of keeping up their skills. Unfortunately, this gift for languages was not passed on to me. Even now that we lived on the same property as Aunt Teresa and I heard short conversations in German almost every day, I never learnt to speak the language, and I didn't impress my teachers with my English abilities either.

In fact they later discovered that I was part of a long line of hereditary dyslexics, not that having a label made things any easier for me. It did however account for why I needed to work so hard if I wanted to keep up with my peers in reading and writing, and yet performed so well in mathematics and science without putting in a lot of effort. I suppose I must have been bright, but with a learning disability which neither my teachers or parents recognised, I did not get a lot of value out of my lessons and certainly never did very well at school. Perhaps worse was that like most dyslexic students I was labeled as unmotivated and so I soon found that it was often easier to go along with this and assumed the role of the class clown.

Each school morning, Aunt Teresa and I would walk out to the cross roads just in front of our property and wait under the 'Finger Post', an old style street sign with 'fingers'

pointing in all the different directions. There was Haigh three quarters of a mile away, to the north west, and Standish beyond that in the same direction. Wigan was shown as 3 miles to the south west, and Boltam to the south east and Horwich to the north east. On top of the fingerpost was an old lantern, with the name Aspull proudly written on two of the four sides of glass. It was the central meeting place in the village and it was here there would be a choice of buses; one from Haigh at five past the hour, one from Horwich at half past and one from Bolton at a quarter to the hour. Throughout my entire childhood and young adult years I don't remember this timetable ever changing. A trivial sentiment, but something which I came to value during the mid 1940s.

Each school morning, on the way to the Finger Post, we would walk past the 'One House'. This was the local council building. It was rather small, (so was the council) but comprised of two storeys. One morning there was great excitement as we passed: it was on fire! The fire engine came from town too late to do anything. Most of the village came to watch and we missed the bus. The cause of the fire was later explained; it was the duty of one of the council workers to arrive first and light the fire. On this cold and frosty morning he apparently thought he would get the fire started quickly and did so with the help of a can of petrol, which was conveniently left in the yard. He started the fire much more effectively than he expected.

It was tremendously exciting, and I was very glad to have missed school. I stood and stared at the red, open seater fire truck, with all its nobs and bells, and the large overhead ladder. It was a beautiful looking vehicle. I don't think I understood the full significance of the situation or its potential consequences. Those were the days when I didn't need to take responsibility for very much. I could just observe, and enjoy the excitement of life. I was the youngest of five children, so I was more sheltered from the responsibilities of life than my siblings had been. My time had not yet come. Although, of course, I was too young to appreciate how wonderful these innocent days were.

Meanwhile, my sister Monica, who was now twenty-one years old, was about to marry and leave home. Monica was a school teacher and she had completed her college training, and obtained a post in Saint Patrick's School, in a less than prosperous part of Wigan. Not that there were many 'prosperous' parts in those days. Monica had met William Charnock a few years earlier, when we had lived in Ormskirk, and he was living on the Liverpool Road, which bordered our property. William was in the agricultural business and how he and Monica met is unknown to me, although I suspect it was in the local choir for they were both keen on singing. Indeed, one of the first things I remember the family listening to, on my brother John's home made wireless radio, was a concert given by the choir in which Monica sang.

William was seven years her senior, and had served in the Great War in the Liverpool Scottish regiment. Liverpool is certainly not in Scotland but I assume there were so many Scots living around Liverpool that it was considered that they deserved a regiment of their own. He had been gassed during the war, which had lasting effects on his health. It was mustard gas, and so the advice for him was to keep exercising his chest capacity, and thus he took up singing.

In September 1927 their much anticipated wedding was held in Aspull. There had been great excitement amongst the family as we counted down towards the grand event. The wedding breakfast was held in our house, but being only seven, I was confined to the kitchen and did not have a place in the dining room among the adults. However I did join the family for a photograph of the assembly. As could be seen for many years, in the family portrait that hung in our house, my socks were in their habitual place round my ankles. Presumably in the excitement of the proceedings no one had noticed and so I had not been given the customary admonition. I was embarrassed about this every time I saw the photo, and whenever I thought of their wedding. From then on, I did my best to dress properly (and somehow over the years, the photo has since disappeared anyway).

Their ceremony was held at a church with the curious name, the Church of Our Lady of the Immaculate Conception. Surprisingly, given my last performance at a wedding, I was allowed to attend, but only on the condition that I did not utter a word. I did my very best and sat straight and quiet throughout the service, and the whole ceremony was without incident.

Johnson Family Picnic C.1933.
From left to right: James Johnson, William Charnock, John Charnock, Judith Charnock, Monica Charnock, Elizabeth Johnson, Christina Charnock, Frankie Johnson, Teresa Moore, Aida Johnson, Cyril Johnson.

Billy Charnock had obtained a post at a Farm School in Birkdale, which was a correctional institution for young offenders. After the wedding, the happy couple took the family dog, Flossie, with them to the farm, which was upsetting for me. Flossie was a big Airedale and she was rather smart although she had not been trained to do anything useful. One day William came home feeling tired and motioning with his arm, he jokingly told Flossie to go and bring the cows in. To his surprise she did so. It was believed she had watched the cows being brought in and so she knew how to do it. On another occasion a sow had a litter more numerous than she could look after. Flossie took it upon herself to look after a couple of little piglets. She laid down, beside the pig and the few piglets that couldn't find a place to park their snout, and squirmed over towards Flossie. She kept them warm, and soon someone put a few milk bottles in the appropriate places.

I missed Monica, and although she lived in a different town, and now attended a different church and village shops, I still saw her quite often. Once we even arranged to have a family picnic, where we met in a park off the side of the road that ran between Monica's new place in Wigan and our house in Aspull. Nearly the entire family managed to attend. Whether it was in the form of letters, telephone calls or face to face contacts, whenever we heard from Monica it was obvious that she, Billy and the dog were all doing well. I don't think Monica or any of us could ever have imagined the consequence of her absence, and in particular, the consequence of us having a spare bedroom in the house.

Chapter Four

Shadows from the Family Tree

FOR EVERY FAMILY I SUPPOSE there are times of crisis when the 'even tenor of their ways' is changed and the turning point for us was the end of 1927. It is surprising how the most well-meaning actions can have the most unexpected and ill-deserved effects.

Going back, my paternal grandfather, John Johnson, had a brother Joseph Johnson who was born in 1828. In the mid 19th century, Joseph had a carrier business, mainly taking coal to Liverpool via the canals and rivers from the mine known as Johnson Pit. My father never spoke much about this uncle Joseph, but I understand that in 1860 or thereabouts Joseph migrated to America taking with him more than his fair share of the proceeds from the sale of the coal mining business in Aspull. Whether or not Joseph left in a hurry is not known but he did leave his watch behind (which must have been well made for it still works and is now in the safekeeping of my grandson). In America Joseph started a very prosperous salvage business in Ohio, which was greatly boosted by the American Civil war and by all accounts he and his family lived an extravagant, high society lifestyle.

However not all was as glamorous as it seemed, for in 1878, two of Joseph's daughters, Eliza Agnes and Mary Emma Johnson, left their American home to return to their extended family in England, although the circumstances surrounding this decision were never revealed. Mary Emma Johnson, known as Emma, was born in 1861, which made her just seventeen when she left

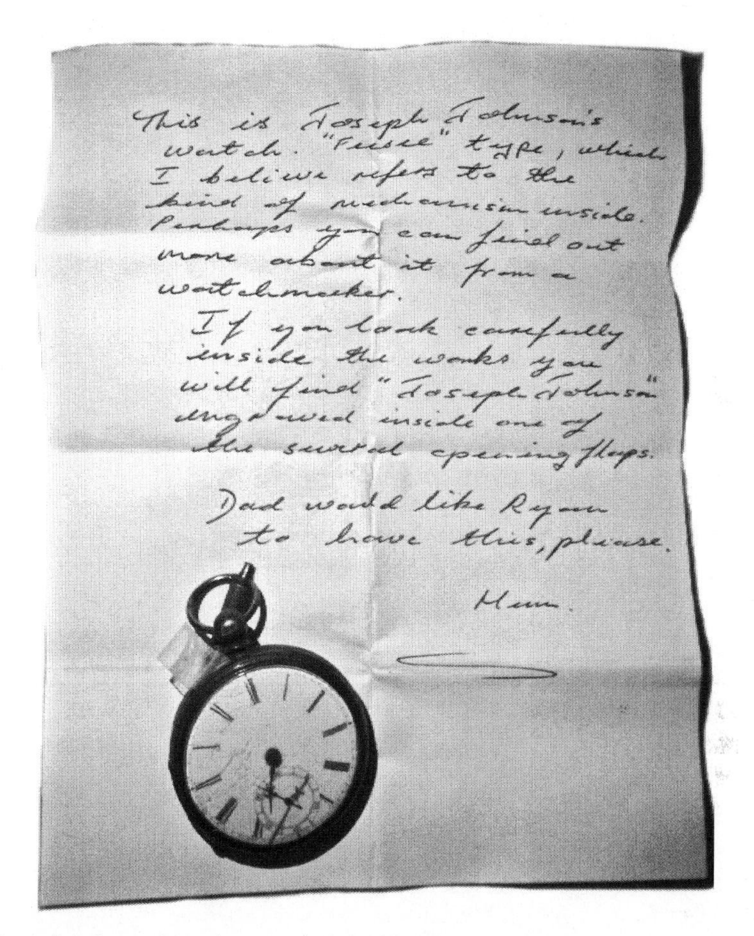

This is Joseph Johnson's watch. "Fusee" type, which I believe refers to the kind of mechanism inside. Perhaps you can find out more about it from a watchmaker.

If you look carefully inside the works you will find "Joseph Johnson" engraved inside one of the several opening flaps.

Dad would like Ryan to have this, please.

Mum.

Joseph Johnson's pocket watch, left behind in England c.1860. It still works more than 150 years later, and has become a family heirloom, left to Cyril's grandson, Ryan Johnson.

America. Despite her youth, and the rarity of two young women travelling alone, Emma and her sister managed to travel from Ohio to New York, board a ship and cross the Atlantic Ocean to arrive in England, at 29 Dickonson Street, Wigan, the home of their uncle and my grandfather, John Johnson, the chemist. My father, James Johnson, who was born in 1872, was only 6 years old at the time.

James was unhappy with the new circumstances in his life and he resented his cousins' influence in the home. With the passing of James' mother around the same year, James went to live with his uncle Thomas, who was also a chemist. James never returned to his father's home, in part because he got on so poorly with cousin Emma, and because when his father remarried, the new Lady of the house and Emma both encouraged one another in their poor treatment of the young boy. Unable, or unwilling, to move back into his father's home, James was taken out of Wigan Grammar School and sent away to a boarding school in France. Although this was a difficult transition for him, he considered having the English Channel between him and his home somewhat of a relief. Some measure of the atmosphere at 29 Dickonson Street can be imagined from the census records of 1881 when James Johnson was recorded as 'domestic servant', and then crossed out and 'no O' substituted for his occupation. He was then only nine years old, and it is little wonder that

Page 40] The undermentioned Houses are situate within the Boundaries of the

Civil Parish [or Township] of	Municipal Borough of	Municipal Ward of	Parliamentary Borough of	Town or Village or Hamlet of	Urban Sanitary District of	Rural Sanitary District of	Ecclesiastical Parish or District of
Wigan	Wigan	Swinley	Wigan	Wigan	Wigan		All Saints

No. of Schedule	ROAD, STREET, &c., and No. or NAME of HOUSE	HOUSES Inhabited	NAME and Surname of each Person	RELATION to Head of Family	CONDITION as to Marriage	AGE last Birthday of Males	AGE last Birthday of Females	Rank, Profession, or OCCUPATION	WHERE BORN	(1) Deaf-and-Dumb (2) Blind (3) Imbecile or Idiot (4) Lunatic
			Robert Allen	Son	Unm	5 mo			Lan Wigan	
			Kate [illegible]	Serv	Do		14	Servant	[illegible]	
215	25 Do	1	Annie [illegible]	Head	Do		40	House Keeper	Do Liverpool	
			[illegible]	Daur	[illegible]		16		Do Wigan	
			Annie [illegible]	Do	Do		14		Do Do	
			Elizabeth [illegible]	Do	Do		11		Do Do	
			[illegible]	Boarder	Do	38		Master Bobbin Turner	Do Do	
216	27 Do	1	Thomas Byrom	Head	[illegible]	81		Justice of Peace & Alderman	Lan Ashton	
			Grace Byrom	Wife	Do		62	Not in business	Do [illegible]	
			Elizabeth [illegible]	Serv	[illegible]		51	Domestic Servant	Do [illegible]	
217	29 Do	1	Thomas Johnson	Head	Widr	42		Chemist & Druggist	Do Hindley	
			Mary Johnson	Sister	Un		39	Dressmaker	Do [illegible]	
			Emma Johnson	Niece	Do		19	Scholar	[illegible] America	
			Mary Jane [illegible]	Servt	Do		[illegible]	Domestic Servant	[illegible]	
			James Johnson	Nephew	Do	9		Domestic Servant	Lan Wigan	
218	31 Do	1	Thomas Hall	Head	[illegible]	60		Provision Merchant	Ireland	
			Bessy Hall	Wife	Do		45		Ireland	
			Margaret Hall	Daur	[illegible]		20	Scholar	England	
			[illegible] Hall	Do	Do		10	"	Do	
			Ada Hall	Do	Do		9	"	Do	
			Alice Hall	Do	Do		9	"	Do	
219	33 Do	1	James Wood	Head	[illegible]	34		General Practitioner Surgeon	Lan Wigan	
			Jane [illegible]	Housekeeper	Do		39	Housekeeper	Lan [illegible]	
	Total of Houses... 5			Total of Males and Females...		6	14			

NOTE.—Draw the pen through such of the words of the headings as are inappropriate.

PUBLIC RECORD OFFICE | Reference:- R.G. 11/3769. | COPYRIGHT PHOTOGRAPH – NOT TO BE REPRODUCED PHOTOGRAPHICALLY WITHOUT PERMISSION OF THE PUBLIC RECORD OFFICE, LONDON

The 1881 Wigan census record, with James Johnson (the 15th row) described as a 'domestic servant', which was then crossed out and 'no O' substituted for his occupation.

he became a man with such a tough exterior and a soft heart reserved only for his wife and children.

After finishing school in France, my father, James worked for a time in his Uncle Thomas's chemist shop where he learned the business of apothecary—the once common term for preparing and dispensing medicines. Emma and her sister, Agnes, were of a religious nature, or so it seemed, and during James' schooling years, Agnes had left 29 Dickonson Street to become a nun. Emma, however, stayed on and became the virtual head of the family, so it was no surprise that when James felt that he had intruded upon his uncle's hospitality for long enough and still could not return to his father's house, he looked for a way out—as far out as possible, for he went to sea.

My father's first voyage was in a large old-fashioned wooden sailing ship and lasted three years. He must have covered an incredible area of the earth's oceans, but even this long absence did not change his mind about his home life. The only thing I ever remember him saying about his return home was regarding the cat. My father was always fond of animals, especially cats, and there was one particular pet of his, which took residence in his uncle's chemist shop. Before he had left, as a regular part of his day's work in the shop, my father would invite the animal to jump on the scales and be weighed. I haven't the faintest idea how, or if, he cleaned them before weighing more medicines. When father returned from his three-year sea voyage he went to the shop, and seeing the cat, he gave the command. The cat obligingly jumped onto the scales. Strangely, the cat had not performed this trick for anyone else while he had been away, but it certainly remembered my father and his voice, and this must have been a nice welcome home for him. From then on he always had a great respect for animals, and taught me to treat them well and value their loyalty.

Father would regularly go to sea, often for many years at a time. In between his lengthy voyages, when he was back in England, he would spend a lot of time with his sister, Martha Johnson, who was a dressmaker. One day, in 1902 while he hung about his sister's establishment, a lady by the name of Aida Monica Mary Cooke, walked in to order a dress. She and James fancied one another straight away, and it was not long before the two were married. The Cookes and Johnsons were, in fact, second cousins if one looks back far enough on the family tree, but this was the first time they had met, and marrying a second cousin was pretty common in those days.

(Mary) Emma Johnson – 'Aunt Emma'.

My father's uncle Thomas Johnson, who had raised him, must have had a very successful chemist business, because when he retired he left his large residence in Wigan and bought a very large four storey house in West Kirby on the Wirral Peninsular. Emma, the daughter of Joseph Johnson, eventually went to live with her Uncle Thomas. Why this bachelor, Thomas Johnson, and his niece, Emma, a young maiden lady, lived in such a sumptuous establishment I have no idea. The house had eight bedrooms, and they were all very large and there was even a long narrow room near the kitchen which Emma called the butler's pantry, although I am not aware of there ever being a butler.

In 1915 Thomas Johnson died. I presume it was about that time that Emma employed a companion, Mrs Baron, who soon became unwell and died, leaving Emma alone again. When Mrs Baron had become ill, Emma wanted some companionship and, knowing her, I imagine she wanted that someone to be useful. So, it befell the lot of my sister Elizabeth to go and keep her company. Elizabeth was 10 years old at the time and I believe she attended the local school in addition to keeping Emma company. When Mrs Baron died, Elizabeth was forced to view her body—something that greatly distressed her. Not surprisingly, Elizabeth resented the whole arrangement, which fortunately ended when she was sent away to Bellerieve Convent boarding school near Liverpool.

In 1927, Emma decided that she could no longer live alone at West Kirby, not that she was alone much, if at all, because someone was always being drafted to keep her company. She decided it was time for her to retire to an institution where she could be looked after. To enable her to search for an appropriate place, my brother, also named John, was given the task of collecting her in the Buick and driving her to inspect the various institutions that had been recommended. None of the places visited seemed to meet the requirements. To make it a little easier for John, who had to drive about 50 miles to West Kirby to collect Emma each weekend, my mother invited Emma to stay with us for a fortnight while the search went on. I can imagine the misgivings of my father but he always did anything to please his wife, and after all, it was only for two weeks.

Emma was unable to find a suitable place, and so with each passing weekend of John's chauffeur driven tours in the Buick, she suggested that it was less and less necessary for John to give up his entire weekends, and that perhaps they should look at just one convalescent home a fortnight. It soon became one a month, and eventually she saw no need for the search.

Emma, in the manner to which we all eventually became accustomed, persuaded mother to make room for her 'few small sticks'. Mother got rid of much of her own furniture to accommodate Emma and her belongings and in a fit of unusual gratitude, which should have aroused some suspicion, Emma said that her worldly goods would be ours when she died. If my father said anything no-one heard it. We soon had a crowded house, which boasted two grandfather clocks and two pianos, one upright and the other a grand and all the additional furniture that comes from a large four-storey home. Her furniture kept arriving and included entire bedroom suites, so my mother kept throwing out our furniture to make more room for Emma's.

By 1944, her invitation for a two-week visit had been overstayed by seventeen years. Even then, the decision to move out had not been due to Emma's kind will, or the thought that my elderly mother may not wish to wait on a lady of leisure any longer. Rather, Emma only moved out after my mother's sister Teresa finally summonsed the courage to tell her to go and stay in a hotel where people would at least be paid to wait on her! In the end, a chest of drawers from a bedroom suite was the only piece of furniture which Emma did not take with her or sell when she did eventually move out, and for all her talk about pharmaceutical shares in the American stock exchange, we never saw a penny (I believe she left a great fortune to another relative or charity). I never understood why Emma was so mean as to take and sell all her furniture when she was moving into a hotel, and leave my mother with a near empty house and no money, when times had become so obviously tough for my mother.

Before things got to that point, our entire family was turned upside down. Emma developed a set routine, which for a lady in her late sixties, was quite amazing. She had her breakfast at some leisurely hour after the rest of us. Then she would play her piano for an hour or so and fortunately she was a good pianist. Following this she would sit at her desk and write letters or tend to her shares before having lunch with us and after that she would go out for the rest of the afternoon. Some days she would visit a family called Lowes, who owned the biggest shop in Wigan, and there she would play bridge or just go shopping for the afternoon. On other days she would practice playing the organ in Saint John's Church, and if I didn't get out of the house early enough then I would be enlisted to go and pump the bellows of the pipe organ all afternoon.

Then there were her visits to Liverpool, and other regular trips to Southport. Emma would often go and impose on the Fishwicks (father's nephews) or anybody else, for short overnight trips and when she did, she would require someone to escort her. With my sisters usually away at boarding school, this task frequently fell on me. As an eight year old boy, I would haul her suit case down to the bus stop, ride with her, then drag the suit case on to the ferry and cross the River Mersey. After that, it was a decent walk up to the Fishwick house and by that stage her case weighed a ton. When we arrived there, she would give me two return tickets, and ensure that I would be back in a day or so to collect her. I never enjoyed catching the ferry back by myself, and found the whole experience rather daunting, but at least we would have a night or two as a family.

Everyone always breathed a sigh of relief when Emma was away. Her presence was a source of constant, but subtle tension. She would offer her opinion on how we fell short of certain dress standards, or inform us in her supercilious tone, that we were 'rude, crude and unruly'. She frequently told us how superior her American upbringing and manners were; although I don't think she ever equated good manners with not outstaying one's welcome, or a willingness to assist with housework. After mother spoke to her, she was diligent about washing her own teacup and saucer, yet this was about the limit of her contributions. To me, having to listen to the old woman—who claimed she suffered from indigestion—belching up for an hour or so each night as she sat in bed reading with the door open, was the greatest sign of bad manners.

Father was unhappy about her being there, as noted by his constant mutterings about 'the wretched Yank' but he didn't want to interfere with mother's arrangement. My mother had always been very soft when it came to looking after the elderly, as had been shown by the way she still looked after her former nanny in the little cottage on the property. Mother would never think of doing less for a relative, and although she was often exasperated, she believed Emma's reports about the unsuitability of the other retirement homes, and could not bring herself to force Emma into a place to which she so strongly objected. Interestingly, I never heard my parents arguing over the matter, it was just quietly accepted as 'the right thing to do'.

So far as the children were concerned, we often expressed our opinions, and whilst we were not allowed to say such unkind things in any other circumstances or about anybody else, the strongest reprimand we got for talking about Emma was 'Shhhh, the woman will hear you' and once we dropped our voices to a whisper, we nearly always got an approving nod or smile out of either parent. Eventually my brother and sisters and I learnt to use 'the woman's' excellent hearing powers to our advantage, while not technically doing anything that the parents could reprimand us for. We would ensure her bedroom door was open and say to one another, that 'she wears a horrible looking hat' or 'she always smells so stuffy', and we could have been talking about anybody, but knowing that Emma was so egocentric, she would think we were talking about her. We had a great time making up people to complain about, when she was just within earshot.

Despite our nasty games, Emma never seemed to like doing things on her own and was always drafting one child or another to accompany her, even if she were only going out on day trips. With the others away so often I had more than my fair share of these outings. Frequently, it would be a trip to the cinema, which she enjoyed, but from my perspective, we went to all the wrong films and perhaps worst of all was the requirement to play games with her, when it was too cold or wet to go out. My complaints were to no avail until one day I refused to play Bezique any more. It was the first time I had totally refused. When my mother enquired as to the reason, my sister Elizabeth well remembered my reply, which she reported as 'Mummy, she cheats!' For that piece of impertinent truth I was sent to bed without my supper.

Chapter Five
Boarding School

I NEVER LEARNED IF EMMA was the cause of my banishment but I feel certain such was the case for at the age of ten, in 1930, I was sent to boarding school. Selected for me was Thornleigh Salesian College, a Catholic school in Bolton, Greater Manchester. It must have been a desperate measure because the first form in the school was for eleven year olds. While most forms had about thirty pupils, a special form was created which contained three of us boys who were either under age or backwards or both: a little Belgian boy, an aggressive Irish boy and myself.

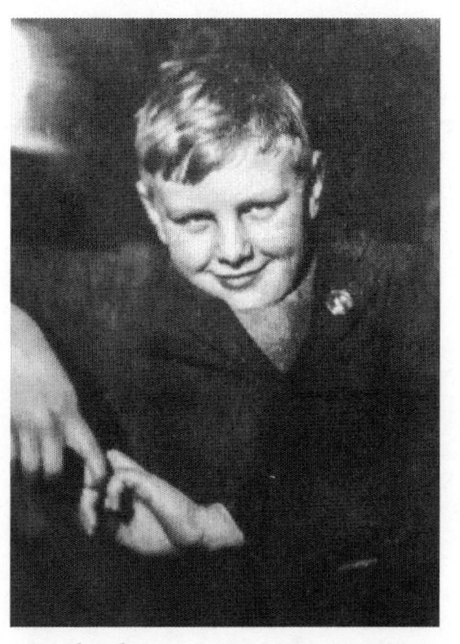

Cyril Johnson, age 10 years, 1930.

From the moment I was first told that these were the plans for me, I was petrified. I quite literally just stood there, unable to move, as my father continued his speech, saying something about opportunities. A heavy weight came over me. It was a strange feeling, unlike anything I had ever known before. The heaviness settled on my shoulders and chest, and no matter what I did, it wouldn't leave. The idea of leaving home for months at a time, and living in an institution I had never even seen, made me feel sick in my stomach. I wanted to say 'I just can't take it', but I knew it would not have done any good. The arrangements had

The Thornleigh Salesian College entrance.
Adapted from an image in 'Thornleigh 1925-1975' Anniversary Booklet.

been made, father had spoken his final word on the matter, and nothing was going to change the situation.

The day before school started remains clear in my mind. My trunk had been packed with the requisite number of starched white shirts, socks, bed sheets and various necessities including toiletries and of course, shoe polish and brushes. Neatly folded on top where they would not be crushed, were my grey uniform shorts and a brown blazer jacket, which in any other context would have looked very smart, but with the school badge picked out in yellow across the breast pocket, I wasn't so sure. To complete the outfit, there was a little brown and yellow tie, slipped back inside its packaging. Stitched inside each piece of my uniform were Cash's personalised fabric labels, each with my name sown in red letters.

Before my trunk was closed and sealed for the last time, mother slipped in a little tuck box with some treats, and home made biscuits. I hoped that this was allowed and would not cause me any trouble. Without even expressing this concern, she looked at me and assured me that every child at boarding school took a tuck box with special treats and reminders of home cooking. I looked at her and knew I would miss her terribly, but before I had a chance to look away, she gave me a big hug, which made it even harder for me to hold my tears inside. One or two spilled.

I came out into the kitchen and ate my lunch, which felt

more like my 'last supper'. It was the usual boiled egg and salad, but after that it would be many years before I could enjoy a boiled egg again; the association overcame any feelings of hunger. The fateful hour arrived when Freddy Wood's taxi drew up at our door. I was too frightened to cry and in any case it would have been to no avail; my fate was sealed.

Father accompanied me on my first journey to Thornleigh Salesian College, for which I was very grateful. In the taxi he tried his best to comfort me, explaining that I would soon make

Surname JOHNSON Other Names Cyril Alphonsus Bede
Father or Guardian James Johnson Occupation Dispenser
Postal Address Holly Villa, Haigh Rd., Aspull { in the County of Lancs / in the Borough of Wigan
Scholarship Admission No. 227
Date of Birth 16. 1. 20. Date of Admission 7. 5. 30.
House Offices held
Position on Admission—Form II C. Position on Leaving—Form VA
Places of Previous Education 19 Notre Dame Convent Wigan
19
Date of Leaving 16. 7. 37.

S 5371 THE INTERNATIONAL STUDENT'S SUPPLY ASSOCIATION.

Thornleigh College Admission Card.

new friends, that I would really enjoy myself and that it was a privilege to be educated like this. There was even a rare father and son moment in which he shared how vulnerable he had felt on his way to boarding school in France, but I was too terrified to take much of it in. I just sat there, staring blankly out the window and nodding; I did everything I could to look brave, as I knew how much this opportunity meant to my father. Before long, the taxi crunched up the winding gravel drive to the very imposing front door of the large ivy covered schoolhouse.

We were ushered into the hall which was impressive with its timber panelled walls and enormous grand staircase. A cricket bat leaning against one of the columns of display cabinets caught father's eye and he expressed his hope that I would learn to enjoy the game. I looked at the wooden bat and realised that, other than my father who would soon leave, this bat was the only thing in the room that was not boastful or overtly ceremonial. Everything else was highly polished and positioned to make an unmistakable impression of excellence, and with that, expectations. The Head Master asked my father to fill in some paperwork, and then called another boy, Thouard who was either French or Belgian, and told him to show me over the school, while he had a word with my father. The only part of my conducted tour that I remember was a dense thicket on the other side of the sports field, called 'Doom Valley'—with a suitably bloodthirsty explanation from my young guide.

The rest of the day was spent with a teacher showing me where I was to eat, the dormitory where I was to sleep and the layout of all the classrooms, with the exception of mine. There was no provision for a class of boys younger than eleven. So, my age was still creating a problem for the authorities even at this stage of my enrollment.

In England the school year begins in September when the sun sets late, so it was still daylight when we were sent to bed. I could see the layout of our dormitory which would eventually become familiar; two rows of four single beds, and at the end was a curtained off section where one of the masters would sit and read or even sleep, but with the curtains extending to the floor it was difficult to tell if the watchful guards were there or not, so I dared not say a word. I looked at the two paintings on the wall, which offered no comfort or imaginative escape and were so unimpressive that they left no lasting memory despite the countless times I saw them. On that first night, I lay on my white sheets and white pillow slip under a light brown scratchy blanket that matched all the other beds. Mercifully the mattress was comfortable, but I was aware of the heavy feeling in my chest, and wondered how I was ever to settle in this foreign place. Through the window I could see a tree close by with leaves that had begun to turn yellow and on a branch a blackbird sang a sad tune, which matched my mood, and so I cried myself to sleep.

It took many mistakes and corrections for me to learn

the routine of the school. From the first morning onwards we were roused at some early hour and after washing and dressing, which included polishing our shoes, making our beds and cleaning our bed spaces, we were then lined up and marched in two lines to an old chapel for the morning service. After that we had breakfast, which included the worst porridge imaginable, with burnt bits distributed in small chunks throughout. Then we would be marched down, more or less in two lines, to the school playground where we were free, until the bell rang to summons us to the cloisters; where we would line up again, in two very neat files, according to our forms and form masters. We marched to our classrooms in silence.

In the beginning, it was all so foreign to me, and I was very aware that I was in a place with many unwritten rules and expectations, most of which I did not yet know. I wasn't even sure what the consequences were for doing something wrong, but I could tell from the manner in which the senior students conducted themselves when the teachers were within ear shot, and the way everybody did things so uniformly, that the punishments must have been serious. It was terrifying. I had three goals: to survive, to avoid punishment and to not make a fool out of myself. I may not have known what all the traditions and expectations were, but I was in no doubt that they were ridged and that I had better learn them very quickly, or my knuckles would surely suffer the consequences.

On the first morning, scheduled for one of the early classes was physical training. Three nervous ten year old boys were sent with all the new and equally nervous eleven year olds. We changed into our 'P.T. kits' and assembled on the grass. The teacher's voice boomed as he explained the rules of the game. As I was anxiously listening, I realised that I had forgotten to go to the toilet, and suddenly I had an urgent need. The teacher appeared strict with his loud instructions and militant pointing around the field. Fittingly, all of us students were lined up smartly and standing at attention. I didn't know what to do. Nobody had explained to me that I only needed to raise my hand and ask to be excused for a moment. I stood there, with my attention half focused on the man's instructions, and half on my bladder. I was so frightened of missing any of the rules that I determined that I would block out the urgent message and focus solely on the briefing. Gradually at first, and then suddenly, the inevitable happened. As I realised what I had done, a feeling of shame and horror came over me. I stood there terrified, now even more unsure of what to do.

The teacher soon pointed at me, or a little lower, and ordered me to go the bathroom. When I arrived, glad to be out of sight, I stood trembling and feeling sick in my stomach. One glance in the mirror revealed that the white material had not provided any disguise. Once I had dealt with the immediate situation and changed back into my normal uniform, I paced up and down wondering if I would be in trouble for taking so long, but too ashamed to go back out. I didn't know what to do with myself, half wishing the time would go fast, so I wouldn't have to go back out there, but trying to savor every moment before the boys came back to the change rooms. I thought of my parents and what they would think. I didn't even know what to do with the dirty laundry. My head was a mess of darting thoughts, my heart beat fast and my cheeks were still flushed.

By the time the class returned to the change room, I had prepared myself to be laughed at. To my surprise, not one of the other boys mentioned it or ever teased me about it. Despite their brave faces, I think there was a mutual understanding that we were all terrified and perhaps they realised that this could happen to any of us, day or night. Word mustn't have gotten out to the seniors or things would have been very different for me.

As time went on, my fear was replaced by loneliness, but I learnt to smile anyway and look like all the other boys, who I suspect were also putting on brave faces. Before long I had a secret inner world of homesickness, covered by a front of confidence and joking. Soon after that, I learnt to separate myself from almost any emotion, and master a stiff upper lip as a part of my British education. We were all similar in this way, and after the first few days, nobody spoke of homesickness or those that did were bullied and quickly learnt to keep it quiet. Thanks to our stoic teachers, who had all lived through World War I, and to the senior students we copied, within weeks, we grew up

The science laboratories at Thornleigh College, 1933.
Adapted from an image in 'Thornleigh 1925-1975' Anniversary Booklet.

and presented as brave, confident and independent adults. Never mind that we were only ten or eleven years old, strutting about in our prepubescent schoolboy shorts; we looked confident and self-reliant and were able to live away from our parents for months at a time.

Each of us developed our own means of avoiding trouble, and this typically meant pleasing the authorities, through academic or sporting achievements, or in some cases being popular and if all else failed, through clowning about. I fell into the latter category.

If anybody had been completely honest, I expect our secret feelings would have all been similar, but I will speak only for myself. I felt trapped; I could not bear to disappoint my parents so I had to make it work, but everything in me wanted to give up and go home. My struggles were magnified each time I compared myself to my parents who had both been to boarding schools, and faced greater challenges with entirely new languages, but they had not complained or made it sound difficult. Likewise, I never heard my brother or sisters complain about it either, so I felt that there must have been something wrong with me to struggle so much. Above all else I felt that I had been gotten rid of and deserted. I deeply regretted not playing games with Aunt Emma, and wondered if that would have changed or at least postponed my fate.

Looking back, I think the biggest problem was that I couldn't reconcile the 'privilege and opportunity' speeches I kept hearing with my feelings of resentment and ungratefulness, and the belief that my friends in the village school, living with their families, were better off. For these thoughts I felt very guilty as I had always been taught to be appreciative, but try as I may, I was not. Fortunately I was not equipped with such an emotional vocabulary at that stage, so apart from identifying a feeling of sadness, or happiness, I did not dwell on all that was in between. All I was aware of was a knot of emotions that made me feel sick in my stomach, so I soon learnt to avoid feelings altogether and focus on what I was there for, to become a highly functional independent adult, able to mix in the right societies, whose relationships and ability to connect with loved ones would somehow just fall into place later on, with any luck. Although the cynicism may have been an optional extra, I think we all developed our fair share!

Beyond this, I learned very little from an academic point of view but quite a lot that was not on the syllabus. I learned what injustice was. If some pupils in a class did something wrong and the master could not find the culprit he would punish the whole class. My view was that if the master was not smart enough to find the culprit it was unfair to punish the innocent. I also learned to keep such opinions to myself.

I got a thick ear every so often and at first I thought I probably deserved these, but to my surprise I also learnt that some of the teachers would just clip us for no reason at all. One master in particular would walk past our single desk files and look at our work, and then occasionally just hit us hard over the ear. It would really flatten the poor boy that copped it, no doubt about it. I remember at least two occasions where I was actually doing my work and the teacher walked by and backhanded me. I am not sure if he thought I was not doing my work or if I had made mistakes greater than my normal spelling errors or if he was just making up for previous missed opportunities, but the seeming randomness of it certainly kept me alert. I soon learnt to be scratching my hair on the appropriate side whenever they would file past the desks.

Other than the usual clip on the ear, punishment in more serious cases involved being sent to the headmaster who used a cane on the hand. If he ran out of hands it was then on the bottom. Anyone who has not experienced trying to sit on hard wooden chairs after this cannot imagine it. Sometimes, however, punishment was just a matter of writing 'lines'. In the lower forms they would be something like 'I must not talk in class'. Doing that a couple of hundred times did nothing for the quality of handwriting. As we became older the lines could be in Latin from say, Pliny or Caesar. We often included a lot of rubbish in copying these, on the grounds that no sensible person would bother to check their accuracy. This assumption was found to be untrue on occasion, with dire consequences.

When we were not being punished, in the ordinary course of the day, there were a lot of aspects that I found difficult, not that I can blame anyone for this. Unlike my parents, I was not blessed with a gift for languages, not even for English but we were expected to learn French and Latin. French was never my forté and I found it near impossible to master all the irregular verbs. My papers would be returned with about as much red ink on them as I had blue.

Thinking about how difficult it must have been for my father to complete his education in this strange language, did little to encourage me, but the thought of his reaction if I failed the subject, was enough for me to focus and eventually sort myself out. Latin was not quite as difficult, and I soon learnt that it was very useful in the interesting subjects like science, which I seemed to do rather well in. English, although not easy for me, was actually very enjoyable.

Of all those who tried to teach me, who do I remember with anything like affection? Only one teacher comes to my mind—a man whose clothes were untidy and stained with snuff, a man who offered me a cigarette and a glass of wine in his study —the most dreadful of sins in the school community—a man whose love of literature outweighed all other faults. For him the characters in Shakespeare's plays were more real than the insignificant people who ruled him. He had a scar above one eye, obtained by playing rugby in his youth, and always looked gruff and not the least like any of the other teachers, but he had a way of bringing the class to life.

The passionate teacher would make each student read a few pages aloud and narrate as though we were professional actors. He taught us effective ways of storytelling and making the most of vocal animation. From my early days in the village school, learning about ancient Greece, I had always enjoyed well narrated stories, and now I was learning to do this myself. His enthusiasm was infectious, and from him I developed a fondness for Shakespeare, and was chosen to play Olivia in "Twelfth Night", a thoroughly enjoyable production. He made a life long impact, and shaped my interest in his subject, English Literature. My best school memories I owe to this man, an otherwise unknown and unassuming man, whose name was simply—Davies.

Cyril Johnson (fourth from the left) dressed as Olivia in Shakespeare's Twelfth Night.

In addition to my ever increasing enjoyment of English literature, I discovered that I had a similar interest in history. Although the historical accounts were not as brilliantly articulated as Shakespeare's plays, when I applied the skills Davies had taught me, and used my imagination to read between the lines, I could see the events unfolding as though I was there. The topics chosen were fascinating and focused predominantly on English history from the 1800s onwards. Although it wasn't a significant component of our course, whenever we got the teacher onto the subject of World War I, he became very animated and passionate.

The history teacher, like all of our teachers, had lived through World War I and we soon learnt that we could also get a few of the other teachers to talk about their experiences of the Great War. While some of the boys initially asked questions as a means of procrastinating through a lesson, we found their stories very engaging, and considered them a better use of time than learning multiplication tables or irregular verbs. Some of the teachers refused to ever share on this topic, but the ones that we could pry open would talk pretty freely and did not spare any details. Some parts were spine chilling, and probably not necessary for boys my age, but I suppose they wanted to instill in us the real cost of war. The message got through to me, I did not like the thought of a war at all, it horrified me and I was glad to live in peace time.

Most of the time school lessons were not so interesting, particularly when we had to stay on task. The classrooms were essentially all the same. Two big black boards across the front, and a teacher's desk facing four or five rows of single wooden desks, the type that had lids which lifted up to store school books in below. Most of the desks bore scars of previous students and their pen knives, but I never carved any of the desks. The desks were all on a slight incline, and at the top, was a flat part in which a little ink well would sit. There was always an 'ink monitor', a boy who was drafted to fill each ink well at the beginning of each day. The use of a fountain pen was strictly forbidden because it did not allow the proper writing style, with a thick downwards stroke and thin upwards. This meant that we had to keep dipping our nibs every few words.

On Saturdays and Sundays or isolated holidays we junior children were frequently drafted into 'fagging' for the staff and seniors who played tennis. Anything more boring than running after tennis balls all day was hard to imagine. For the first year we were fortunate in having an easy going and enlightened Rector, the most senior staff member from the church whose authority was higher than the Headmaster's. This Rector would send us with equally enlightened staff to watch the local football team or, if they were playing 'away', then to the cinema.

The cinema I found to be more educational than the school lessons. The films were usually black and white, though I do remember seeing one in colour once, for which we had to wear special glasses. Sound had come into cinema only a couple

The classrooms in Thornleigh College, 1933.
Adapted from an image in 'Thornleigh 1925-1975' Anniversary Booklet.

of years earlier, so this was still very exciting. We saw wonderful action films suitable for boys, such as 'Charlie Chan Carries On', and 'All Quiet on the Western Front' and from time to time a beautiful actress would feature reminding me of Lilly Moss, of whom I was still very fond. We also saw 'The Dawn Patrol' which was a most inspirational experience in which I felt I was flying the Newport 28, a French biplane. My imagination ran away as I thought about how thrilling it would be to fly in an airplane, and imagined the sense of freedom and wind on my face, but of course eventually the credits came up and I had to come back down to earth, and worse, school. On other occasions we would go for long walks over the moors, often ending in a restaurant for a meal the like of which the school could never provide. We did well for weekend entertainment, for a while, and then we got a new Rector, who we guessed came from the deepest bogs of Ireland, and he sternly put an end to such sinfulness as indulging in cinema.

After that I came up with the idea of making my own motion picture by drawing a small stick figure at the bottom corner of each page of my exercise book, such that I would be able to flip the pages and see the figure come to life. If this worked, I had plans for expanding the plot, but to begin with, I first had to get the main thing right, the flight of a new single engine De Havilland Tiger Moth. So began many little drawings, but it was taking longer than expected because classes were eating into my schedule. Of course, I took it in with me and started working on the important project during lessons, but I was caught. The teacher did not see the importance of this great work, or its relevance to French grammar. I got another thick ear, which would have been worth it, had my book not been taken away.

Cyril playing tennis on the grass courts at his grandfather's estate (1933).

With the new Rector and cancelation of all other weekend entertainment, most of the staff and pupils were devoted to sport and those who were not very interested were regarded as unworthy. We boarders were all made to play football or cricket in house teams. The school was represented in matches against other schools by first and second teams and 'under fourteens'. I had no love of football and I much preferred rugby, a game the school did not play.

During the summer holidays when I was able to return home, I would occasionally play tennis on grandfather's grass courts (the ones which I had not previously destroyed in my attempts to be an ice-skater) and whenever there were a few village boys available I would organise a game of cricket. My father was especially keen on the game and I marveled at his ability to bat or bowl left or right handed. In response to my complaints that I never got much of a chance at school, father put in some intensive coaching. I enjoyed spending time with him, and seeing his willingness to do this made me eager to succeed and master the skills. He provided a single stump that I was to aim for and on the ground he pinned a piece of paper. I was to bounce the ball on the paper and hit the stump. The next step was to move the paper and still hit the stump. When I was able to do that, at least some of the time, he let me increase the speed of delivery.

Triumph came when back at school, we were fielding at the nets with the school's favoured bowlers and batsmen. I caught a ball and instead of throwing it to the bowler I bowled it myself. The batsman, a first team man, never 'ad'a chance', I got him. 'You can't do that again Johnson!' someone yelled, but I did and continued to do so. That put me straight into the first team, which was some sort of victory for me, although apparently not considered so by some of the sporting types. I have often wondered if anyone at Thornleigh has ever beaten my season's average—at the time it was the school record.

Of course I was the last man in the batting order; but when the first team played the second team, with the first team in reverse batting order, I had my revenge against the 'system'. I was first man in and I took no chances, staying to the end, still not out. It did not add to my popularity but it worked wonders for my sense of justice. I couldn't wait to see father's face when I told him next holidays, but I didn't allow myself to be too excited, for I knew that by the time I next saw him, the story would be old news and not mean as much to either of us.

As I had been for most of my life, I was one of the unhealthiest students. For the first few years I never completed a year without at least half a term at home with some complaint such as measles, mumps and suchlike infections. This always seemed to happen when going home for a half term holiday. I also learned that reporting sick at school was something to be avoided, as the treatment was worse than the disease. The treatment for every complaint always seemed to involve taking a terrible brown medicine which seemed to be the only cure

Brother James, the nearest the school had to a medical expert, could provide. Even if we were in a state of good health Brother James would sometimes come around the dormitories at night to dispense his evil brew.

Accidents were all treated in a similar, almost uniform, manner—with iodine and a bandage. I once conducted an experiment, extra curricular of course, with some carbide in an old bottle. The resulting explosion was quite gratifying although it left me with the top of one finger hanging on with a minimum of skin and flesh. I wore Brother James's bandage for a couple of weeks without getting any infection, but the most painful part of the whole incident was getting the bandage off.

On another occasion I came home for the mid-term weekend with festering green sores on both ankles. I explained that I preferred to put up with the considerable pain and discomfort rather than report sick to Brother James. I think it was this incident that made my parents send me to school as a weekly boarder (an apt description). Later, at about 13 years, when my father considered that I would be less likely to kill myself travelling to and from school every day, I became a 'day boy'. Travelling required catching a very early bus and only just getting in before the parade in the cloisters and arriving home at 6 o`clock in the evening. After tea there was homework, which seemed to take me longer than everyone else.

Having weekends at home gave me the opportunity to develop my two hobbies; flying and photography. During my first year of boarding school, on one of the holiday breaks, I visited a secondhand bookstore with my friend Francis Ball, the grandson of the coachman. We had put our money together to buy an old American model aircraft magazine, simply because it looked interesting, and so began a lifelong hobby and passion for two ten year old boys. We spent much of that holiday period and the next few breaks, trying to obtain materials to build similar inspirations with little success, but now that I was home on weekends, I had a lot more time to pursue this passion with Francis.

On another school holiday, I returned home to see my Aunt Teresa cleaning out all of her former husband's belongings.

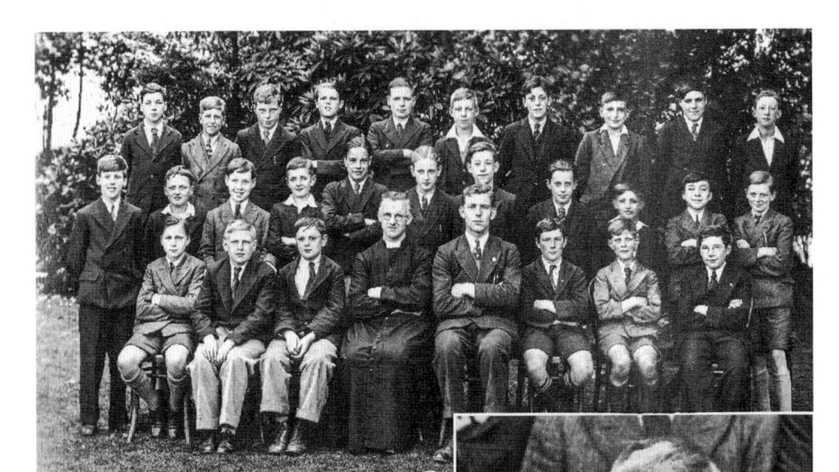

Thornleigh class photo,
Form 3B, 1933.

There was a great pile of his in the laundry, all ready to go to charity, so she invited me to look through it and have anything I fancied. Troggles had been a passionate photographer, who had in fact taken many of the family portraits, so when I came across his old fashioned plate camera, I decided to see if I could do anything with it. It turned out that I was actually very keen on photography, and thanks to Troggels' substantial supply of dark room equipment I was able to develop all my own prints and indulge in this hobby without it costing me anything.

These hobbies distracted me from the schoolwork I should have been doing and unlike the early years when I was building in Bobbie's garage, there was no Sheila Ainsworthy to rescue me.

My final year at school started badly. My father was still, even at my age, a rather stern man who always kept his word and meant what he said. The 'Ship's Captain' aspect of his character could still be seen at home on occasion, particularly with his teenaged children. Father explained to me that after this year I would be leaving school and he did not intend to maintain me in idleness. His graphic descriptions of my future were terrifying and I was duly impressed with the need to mend my ways. Until this moment I had always been quite happy to be the bottom of the class, a position which another boy and I alternately shared. My life changed and I listened in class and *worked*. By the end of the first term I found myself to be the top of my class, and was issued a certificate of merit, declaring in red ink, 'First Place'.

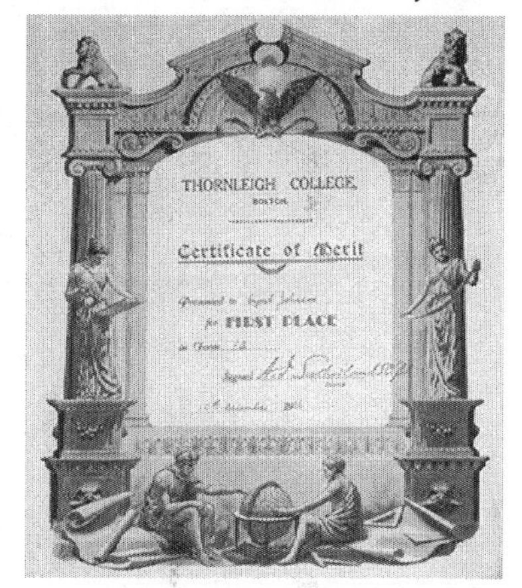
THORNLEIGH COLLEGE

Certificate of Merit

FIRST PLACE

Cyril's certificate of merit, 'First Place', 1936.

This feat did not gain me the adulation I would have expected from my father; rather I was upbraided for having wasted the previous years!

Chapter Six

If I only had Wings

MY INTEREST IN FLYING MACHINES began when I was about ten years old and unlike most phases that young boys go through, it has remained throughout my entire life. My first attempts to make model airplanes were hardly a success, but my friend Francis Ball and I learned a great deal from these experiments. It was only 1930 and the whole model aircraft industry was still in its infancy, so enthusiasts—usually with more money and experience than us—spent most of their time experimenting and doing their best to create aerodynamic models. From what I could see, even amongst the experts, more models crashed than landed, but this did not deter us.

Our first model was based on a picture of a glider, with an open frame that supported a wing on top, a fin, rudder and tail plane at the back and a pilot sitting above a skid. It was supposed to be towed into the air by some athletic man running down a hill. At that stage we had never heard of balsa wood and would have had little means of buying any even if we had. I found a piece of straight grained wood, something that father had brought home from Africa, undoubtedly meant for some other purpose, but which I somehow managed to cut into thin strips. The fuselage was made roughly like the picture and stuck together with a hot glue gun. The wing and tail end presented a challenge, but making wooden rectangular frames and covering them with an old rag, we thought would overcome this. The glider's first and only flight was very instructive; bricks fly in a vertical direction. It was disappointing, but not discouraging.

We needed to find more plans and instruction on the matter. In the local town, down a narrow arcade past the man with a large set of scales who offered to weigh anyone who was weight conscious, there was a second-hand bookstore. Here were used books and magazines of every description including some to do with model aircraft, as we had found during the last holidays when we pooled our money and purchased our first aviation magazine. There were very few of these, which I suppose reflected the lack of interest amongst the population and also probably, as in our case, a lack of money. It was here we also discovered *Flying Aces*, the most beautiful reading material we could imagine. This magazine opened up a whole new world, a world of balsa wood, which sounded exotic. The magazine made promises of it being lightweight, flexible and soft enough to cut with a razor blade. It would be the perfect material, but there was a shortage of tropical South American trees here in Wigan. Unlike the model magazines of today which are little more than catalogues, *Flying Aces* told the reader how to do things and gave drawings of what they could construct. In addition, there was always a story about the life of Phineas Pinkham, a pilot from the USA. Heaven!

The August 1929 cover of *Flying Aces* by Periodical House. Inc.

Eventually we were able to source this incredibly light wood at a store in Liverpool, and made many models, some of which flew, and some did not. We made models of a French Caudron, a Curtis Racer, a Fairey Battle, and some which were a combination of these and designed purely as experimental flying models. Francis made a flying wing. This was circular and although beautifully made I never saw it airborne.

On occasion we would invest in kit models that were very instructive and these gave us new ideas to try on our own creations. One Christmas I was given a Wakefield rubber model; it was large yet very light because it was made of tissue covering fine sticks of balsa, and powered by a wind up rubber band. This performed well with its twin rubber motors geared together

One of Cyril's earliest model aircraft, an experimental hybrid.

to provide considerable power without twisting the fuselage, all of which was very impressive and gave me further ideas for building, but this would have to wait until the next school holidays.

Over the Christmas holidays of 1931, when I was home from school, I made a very detailed model of a Hawker Hind. It was a most beautiful model but required hours of work to construct such details as the Vickers machine gun in the rear cockpit. This was far too delicate and good looking to risk in the air. Rather it took pride of place at the top of my wardrobe; then one day it did fly, thanks to someone leaving the window open on a blustery day. Fortunately it sustained very little damage, and landed safely on my bed. I was so relieved when I examined it and found that nothing had been damaged, not even the undercarriage. I was proud of its construction, but still decided it was best suited as a display only model. I had two other favorite models, both of which were remarkable. One was a model of a Fokker Triplane, by the Cleveland Model and Supply Company. This was so small and light that it was unsuitable for outdoor operations but it glided well in a small room with no draught. Then on fine days I had a more robust model of an aircraft called 'Mister Mulligan' and this performed exceptionally well. We constructed more models than I can remember and a

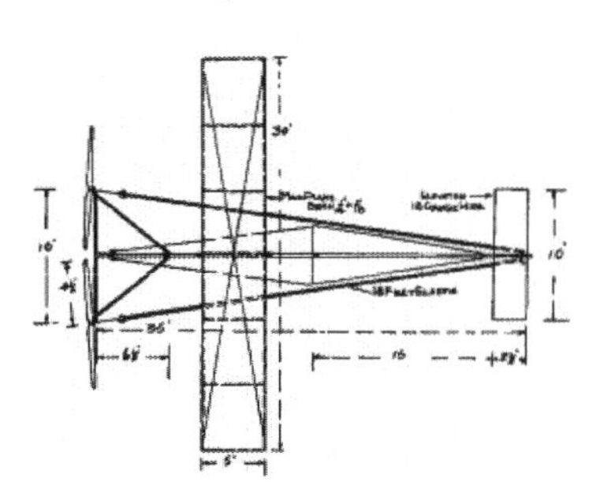
The plans for the 1911 Wakefield winner model, Twinning monoplane.

A large tissue covered Wakefield model, a winning design from the early 1930s, similar to the one Cyril would build and fly.

few summers later, I built and launched a Fairey Battle, which climbed at a surprising rate to quite a height and descended even faster, vertically. A lot of our experiments did not end well but we enjoyed ourselves and continued constructing models in the hope that we would eventually master the art.

Francis and I would try to repair and salvage the material from each model as they suffered their various and often unfortunate landings. Materials for building were hard to come by and eventually there were enough glue and repair joints on our models that the additional weight started to impact their performance and made crashing even more inevitable. We realised that we needed money to fund our important projects, and as there was almost no source of employment in the town, much less for two twelve-year-old boys, we decided to create our own.

The Hawker Hind model hand-made by Cyril Johnson.

Cyril Johnson and Francis Ball launching model aircraft in the garden.

Our idea was to create a little newspaper, called Aspull News, in which we could promote what was happening in our village, and as the other newspapers focused on the larger towns, we had no competition and the potential to do rather well. Our newspaper would include a lost and found section, engagement and wedding announcements and any social activities or dances that were coming up, but of course we had to try to find some news to include too. In this sense it was no different to a modern day newspaper in which the publishers need to find some articles just to keep the ads apart, and we too did not want to give the impression that we were more interested in the ads than the news.

We advertised for our news and told people that we could write articles for them if they wanted to circulate a message. We did rather well with births and deaths, because people liked to read about these, but in such a small population, they were rare. Mostly we wrote only local village articles, such as an apology from a farmer whose cows had wondered out onto the road the week before, someone who had taken ill and would enjoy the company of any visitors or someone who had been reunited with a lost pet. In Aspull—a mining town without a working mine—it was so difficult to find employment that if someone did get a job they would be the talk of the town! Naturally, this made for a wonderful feature story, and the joke was usually to report on how far the new employee would have to cycle to work—often well over an hour in each direction.

We could not obtain a typewriter or anything as fancy as

that, so the newspaper would need to be handwritten. At first this looked as though it would be very time consuming, but we managed to pool our allowances and eventually we obtained a special kit for making copies of paper. In this we were provided with a special ink with which we would write out a page and then melt a jelly substance into a tray. The paper would be placed face down on the jelly, and the ink would come away from the original paper and settle onto the jelly. From there, we could take the next blank sheet of paper and smoothly roll it on, leave it to settle for a second and then carefully peel it off. A small portion of the ink would transfer onto the page making a clear copy and leaving behind enough ink for about twenty more copies. It took a lot of time, but we eventually made around fifty copies of a double sided newspaper, which we folded neatly and then we were ready for distribution.

People tended to receive our entrepreneurship rather well, in part because we were so young, but largely because we gave the newspapers out for free! We thought that this form of distribution would have two advantages, firstly we hoped that with a large enough circulation, eventually someone would pay to advertise with us and secondly, it would enable us to advertise our own services for free. It worked, and within a few editions, we had the local butcher, chemist and two grocers paying us to advertise in Aspull News. Furthermore, a lot of people responded to our own advertisement—window washing, one penny for one window. So we delivered the newspapers with our window washing buckets on the back of our bikes which made us feel rather professional, until we had to explain to anyone with a two storey house, that we were very sorry, but we could only wash the windows on the ground floor because we didn't have a ladder.

Between the advertising fees for our newspaper and our window washing, we made a bit of money, but much of it had to be recycled into the business, to buy more copy ink and paper, so there was little left for all our efforts. Nonetheless, it was growing slowly and we thought that if we held out for a year or so, we would be able to charge more for advertising and have more clients to make it work. Of course school got in the way of some of this, and our circulations were not always regular, but we tried to do them as often as we could, and we persisted for a few years. Eventually we found that the newspaper, although now profitable, was taking up so much time that we scarcely had any left for the flying hobby that we were so desperately trying to fund. We had to reduce the publication frequency until we had just the right balance.

As well as building models, Francis and I were keen to learn about other aspects of aviation—real scale aviation. Over the summer holidays in 1932, whilst still in our early days of learning about flying and the difficulties of smooth landings, we had an idea. We had come across an old sheet, a big one, and decided that it would make a good parachute. Lengths of thick twine were attached to the edges of the sheet and at the appropriate opposing ends to make a sort of harness. Although it was a large double bed sheet, it was doubtful if it could hold the weight of a twelve year old boy. Fortunately for us, if not for our poor guinea pig, there was a small younger boy called Freddie, who used to hang around. He was only eight years old, and we reckoned he would be just the right size for a test pilot. Freddie seemed to be quite keen at first. We explained that the idea was to run and jump off the steep slope, almost a cliff, which was conveniently situated in the field near grandfather's estate. Freddie was a little less enthusiastic after the briefing, but we reassured him that this was, more or less, how the Wright brothers had done it, and a few minutes later he was launched.

Thinking about it later we came to the conclusion that we should have considered the direction of the wind, which was blowing from behind the top of the slope, and thus down the slope. This apparently hastened rather than retarded our test pilot's descent. We helped dust him off to the point that he could limp home, but we did not see Freddie for the next few days and from then on he was much less curious about what we were doing. Fortunately the injuries were not serious enough for parents to be involved, and I was grateful to have avoided the reactions of father or Emma, neither of whom would have understood.

Although the parachute idea had not been a success, I became increasingly focused on the challenge of achieving a man-carrying flight, and a visit to Cobham's Flying Circus inspired me more than anything ever had before; I was determined to be a pilot. In my financial state there was no chance of taking any private flying lessons but there was a glimmer of hope when in 1934, the Navy advertised for boys of about 14 years old, which I was, to join the Fleet's Air Arm. I was very excited and rushed home to put together an application, but my father put an end to any such hopes by pointing out that my physical condition would tell against me and furthermore, my academic record would ensure the Lords of the Admiralty would not employ me even to scrub decks! He advised me that I would be better off focusing on my schooling if I was ever to make a success of myself, and with that advice still ringing in my ears I was sent back for another term.

Meanwhile the prospects of conflict in Europe grew increasingly likely. We read in the newspapers that in Germany the youth were being assisted to learn to fly and it wasn't long until the British Government came to the conclusion that our country was falling behind on this front. As a response by 1938 Sir Kingsley Wood announced the brilliant idea of forming the 'Civil Air Guard'. It was open to anyone over 18 years, this was my chance, I was sure of it. One could join and learn to fly for 2 shillings and 6 pence per hour, which was very cheap. I quickly put my name down and I think everyone else did the same. I was very excited and waited eagerly, but I never heard back from them, and the scheme ended before they could process all the applications. I continued to focus on my education, and somewhat discouraged, I thought the best thing was to study hard and just hope that my results would be of some value in my pursuit.

It was only a few months after that an opportunity arose to join the Royal Air Force Volunteer Reserve (RAFVR). The Reserves were taught to fly on weekends and during a two week training period every summer. My application was received and this time I was sure it would be different. All I was waiting for was the message to report for duty at Barton Hall, the Lancashire RAF Airbase, or so I naively thought. When the anticipated letter did arrive, I was instead instructed to report to the headquarters

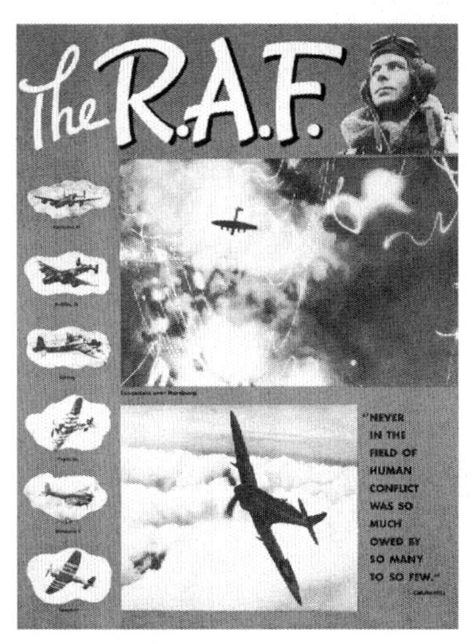

Typical recruitment poster published during the course of the war.

in Manchester for an interview and medical examination, but this did not deter me. This was my chance, I just knew it. I attended full of hope and self-confidence.

The interview was conducted by a row of Royal Air Force (RAF) officers seated behind a long table. First one then the other would ask a question. Suddenly the presiding officer turned over a piece of cardboard on which was written a quadratic equation. 'What does Y equal?' he asked. Fortunately I was able to answer that almost before he had asked the question. His grunt I took to mean that he approved. After some more questions about what games I played and other interests, he suddenly asked, 'If you wanted to fly from here to Paris what course would you steer?' My answer was within 10 degrees and seemed to be adequate, so I was dismissed in an approving way.

Then it was lunchtime. I fell in with a bunch of other young hopefuls and went to the local pub, which served a very good steak and kidney pie for lunch. That, and a pint of beer, improved my self-confidence to a disastrous level. The afternoon was spent in medical tests of every sort. On reflection some of the things they had us do were quite odd. A steel rod about an inch diameter and about 4 inches long was balanced on end on a piece of wood. The test was to pick up the wood from a desk and raise it to shoulder height, without it falling. Then there was the gas meter and one had to blow into that while it measured the capacity of one's lungs. Blowing into a column of mercury and holding the level for as long as possible was easy, as my cousin Peter and I had spent the summer trying to see how long we could hold our breath at the bottom of his swimming pool. I think the Doctor was surprised at my result. The colour blindness test went without a problem, and I was even called back to tell another unbelieving applicant what was on the picture because he could not see the number and he didn't believe the doctor. The final thing was the damned stethoscope. The doctor detected a murmur in my heart and asked if I had ever had rheumatic fever.

The doctor did not say much, but I knew from his face straight away. I was told I could go home now and wait for further correspondence. The atmosphere had palpably changed. It was a beautiful sunny day as I returned home but to me I could have been walking through sleet and sludge and I don't think I would have noticed. I felt despondent and could hardly believe it, even after everything had gone so well in the interview. I wondered what to do next. I tried to take comfort in the thought that just maybe the doctor's findings would not be a disqualifying factor but it was obvious from his face that it would be. Any residual hope was squashed the following week when I received a letter thanking me for my application and informing me that I was 'permanently unfit for all flying categories'.

My uncle Alphie was also a doctor and he had a friend, another doctor, who happened to be the flying club's Chief Instructor at Speke Airfield near Liverpool. I had never sought help from anyone before in my quest, but now I was determined. I went to see the Chief Instructor and asked him what I should do. He suggested that far from having a good lunch and beer, I should have little or nothing to eat for hours before the physical examination. Apart from that he thought it would be very difficult to disguise a heart murmur but advised that general physical fitness was the best solution. After that I played rugby and cycled everywhere rather than use the local bus. I was not giving up. Besides, the Royal Air Force could not really have meant 'permanently unfit' and in 'all flying categories', could they?

Ref:- RAFVR/MAN/249.

Tel:- BLAckfriars 1054.

R.A.F. Volunteer Reserve,
3, Albert Street,
Manchester.

12th May, 1939.

Sir,

With reference to your attendance at the above address on the 10th May, 1939, I am to inform you that you were assessed as Permanently Unfit for all flying categories by the Travelling Medical Board.

In this connection it is regretted that your application for enlistment into the R.A.F.V.R. cannot be considered.

I am, Sir,
Your Obedient Servant,

P Butler

Assistant Commandant,
R.A.F.V.R.
Manchester District.

Mr. C.A.B. Johnson,
Holly Villa,
Haigh Road,
Aspull,
WIGAN.

Cyril's letter of rejection from the Royal Air Force.

Chapter Seven

Learning after School

AFTER WHAT SEEMED LIKE AN eternity the day came when I left school for the last time. I walked through the gates throwing my hat in the air. Freedom! I thought. I waited during that summer for the results of my matriculation examinations to arrive; they did and for a backward and, I must confess, lazy child, the results were not bad. I never regretted leaving my 'alma mater'. I would have cheerfully throttled whoever it was who said that 'school days were the happiest days of one's life'. They certainly were not.

Of course the next big decision was what to do with my life. I had long since abandoned the more youthful ambitions of being a museum curator, the Prime Minister, a lawyer, or even a doctor (and there seemed to be an overabundance of them in the family anyway), but I still harbored a desire to fly. Unfortunately, for the time being that appeared unlikely, and again, the phrase on my RAF rejection letter came to mind. In another dream I saw myself building bridges in darkest Africa and bringing civilisation to

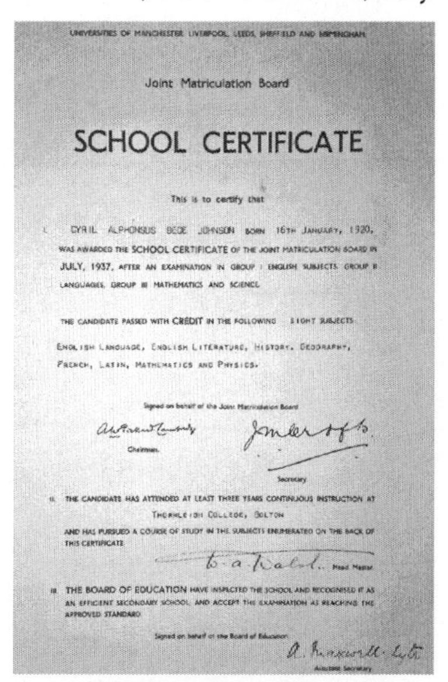
Universities of Manchester, Liverpool, Leeds, Sheffield and Birmingham

Joint Matriculation Board

SCHOOL CERTIFICATE

This is to certify that

I. Cyril Alphonsus Bede Johnson born 16th January, 1920, was awarded the SCHOOL CERTIFICATE of the Joint Matriculation Board in JULY, 1937, after an examination in Group I English Subjects, Group II Languages, Group III Mathematics and Science.

The candidate passed with CREDIT in the following eight subjects:

English Language, English Literature, History, Geography, French, Latin, Mathematics and Physics.

Signed on behalf of the Joint Matriculation Board

Chairman.

Secretary

II. The candidate has attended at least three years continuous instruction at Thornleigh College, Bolton and has pursued a course of study in the subjects enumerated on the back of this certificate.

Head Master

III. THE BOARD OF EDUCATION have inspected the school and recognised it as an efficient secondary school, and accept the examination as reaching the approved standard.

Signed on behalf of the Board of Education

Assistant Secretary

Cyril's certificate from the Matriculation Board, showing that he graduated with credits in all eight of his final year subjects.

the local people, somewhat like I had imagined father doing, but this seemed impossible too. My mother refused to give me any advice as to what career I should follow, on account of her other children complaining that they had been misled in adopting their parent's advice on the subject.

The family fortunes could not extend as far as sending me to Oxford or Cambridge, unlike some of my friends from Thornleigh; but in the nearest town at the 'Wigan District Mining and Technical College' I found that I could do a London University degree in Engineering, and it would not be beyond my parents means. This particular college had a very good reputation around the world and was often asked to recommend its mining and engineering graduates for various positions. The syllabus was extensive, and as a consequence, seemed to require more hard work than would have been required at such prestigious colleges as Oxford or Cambridge, albeit without the accolades. This feeling was confirmed by one of my fellow students who was given a 'Miner's Welfare Scholarship' to Cambridge after completing his intermediate exams.

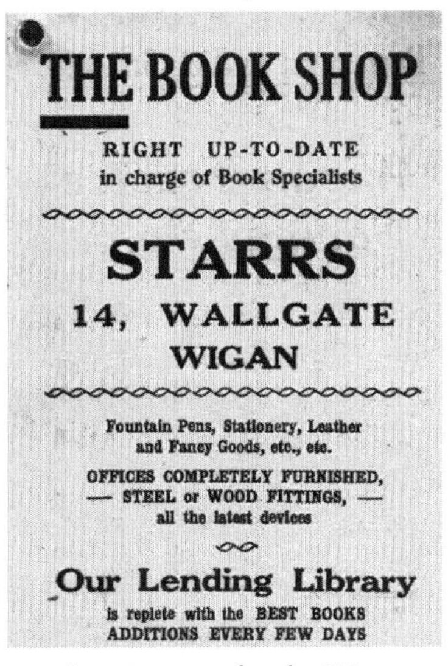

Advertisement for the Wigan bookshop, where Cyril purchased his wooden compass and geometry set, and the necessary entropy and logarithmic tables for his Engineering degree.

Men who had experience in the real world, and who were not lifetime academics, gave lectures at 'The Tech', and so a fascinating world of applied and theoretical physics presented itself to me. An Indian, Mathura Mohan Das, gave the lectures on magnetism and electricity. Other physics lectures were by an American who had previously worked in an astronomical observatory in the U.S.A., a man named Mr Harvey. He gave me a lot of freedom to pursue my own interests and I had the full use of the darkroom that enabled me to continue and expand my interest in photography. About four or five weeks before the end of year exams this friendly Yank pointed out that I had been following my own interests at the expense of the prescribed syllabus. That meant a lot of last minute swatting; however, in the final event I did manage to do rather well.

Soon after, a new problem arose: my father developed pancreatic cancer. It all seemed to happen rather suddenly, or more likely I was only informed of it towards the end. One day I was on holidays at the beach with my friend, and the next moment I received a message to come home immediately. There was nothing Doctor Cooke or 'Uncle Billy' could do and father died very shortly after that.

My siblings and I made sure to keep an eye on mother and to try to do everything we could to help her, but she made it clear that she did not want to lose her independence and seemed to cope remarkably well. I expect this was largely because she was putting on a brave face; but I also think that in part, it may have been because she was so used to father being away at sea for years at a time. I could understand; her reactions and mine were not particularly dissimilar, because I had loved my father from a distance for so many years, while he was at sea, and then for most of the previous eight years while I had been at boarding school. My father had been, and would remain in my mind, a hero who was celebrated in the stories that were told about him, more than the actual times we had managed to spend together. I had succeeded in keeping his voice alive in my mind throughout all the years at boarding school, and during all the times that I had missed him, and in some sense this made the transition easier for me.

As a typical sailor, father had considered that money was meant to be spent and I understand he left my mother with a total of four pounds ten shillings in the bank. By this time, my grandfather Luther Cooke had retired, and through supporting my Aunt Teresa and her children, as well as various others along the way, the safety net of his fortune had eroded. My youngest sister, Frances, was now a qualified pharmacist and had chosen to go to work in London—primarily due to 'the Emma effect' and the subsequent desire to leave the family home—and fortunately she was able to send a portion of her pay home to mother. Ironically, part of the problem was that mother was still supporting Emma. Likewise my brother John was also able to help out a little bit, but it was obvious that continuing with my studies would be beyond our means. I was prepared to look for a job when I was called to see the College Principal. Incredibly, over that same period, I had been awarded the Eckersley Scholarship and Studentship. This meant that not only were my tuition fees paid, but I would also have a little extra cash to help the family finances. Then, because of my interest in physics, the Principal told me that after finishing my Bachelor of Science in Engineering I also had a place waiting for me, with an approved scholarship at Cambridge, to do a Post Graduate course in Physics.

It was not just our family who had fallen on hard times. Since 1930, we had all felt the effects of the Depression. In addition, almost everyone within a five-mile radius of our village had become unemployed when Wigan Coal and Iron Company closed. It had employed a few thousand people, and was the only industry that supported the district, apart from the odd cotton mill. But it was not only the employees of that company who were affected,

DEATH OF CAPTAIN JAMES JOHNSON

Interesting Wigan Associations

Captain James Johnson, of Holly Villa, Haigh-road, Aspull, died, we regret to announce, on Friday last week at the age of sixty-six. His wife is the sister of Dr. R. A. Cooke, Medical Officer of Aspull, and of Dr. W. E. Cooke, the well-known pathologist, of Wigan.

Captain Johnson was formerly in the mercantile marine. A native of Wigan, he served his time, as a boy, with his uncle, the late Mr. Thomas Johnson, who was in business for many years as a chemist in Market Place, Wigan. At the age of nineteen, however, thirsting for adventure, he decided to go to sea. His first voyage was one of three years on a sailing ship. During his early days at sea he was once shipwrecked on the West Coast of Africa. He joined the Elder Dempster Line, and subsequently attained the rank of captain. During the war, he was in the Royal Naval Reserve.

Captain Johnson retired from the sea in 1925. He and his family took up residence at Aspull about fourteen years ago, and he became dispenser for Dr. R. A. Cooke. He was a member of the Knights of St. Columba, Wigan, and was associated with Our Lady's Church, Aspull. In his early years he was a member of the congregation of St. John's Church, Standishgate.

He leaves a widow, two sons and three daughters.

The funeral took place on Monday. A service at Our Lady's Church, Aspull, was followed by the interment at Haigh Cemetery. Fr. J. Holland officiated, assisted by the Rev. O. Fishwick, a nephew of the late Captain Johnson.

The mourners were: Mrs. Johnson (widow), Mr. J. C. Johnson and Mr. C. Johnson (sons), Miss E. C. Johnson and Miss A. F. Johnson (daughters), Mr. W. R. Charnock, representing Mrs. Charnock (a daughter), Miss E. Johnson (cousin), Dr. R. A. Cooke and Mrs. Cooke, Dr. W. E. Cooke and Mrs. Cooke, Dr. Gray, and Mr. Hunter. A large number of friends and parishioners were present.

The bearers were members of the Knights of St. Columba.

Floral tributes were sent by: The Family; Nurse Standish; Nellie, Pollie, Katie, and Martin; Neighbours; Marsden and Balfour; Dr. and Mrs. W. E. Cooke; Dr. and Mrs. R. A. Cooke; Hunter's, Chemists, Market Place, Wigan; Mrs. Armour and the Misses Athron; Mrs. Preston; Mr. Horrocks; and Mrs. H. Moore.

many of the other locally owned mines actually became flooded when the Wigan Coal and Iron Company ceased to operate its underground water 'pumping pit' as there was a shift in the water table. Consequently the steel mills and almost everything else came to a halt.

My brother John, who was very good with his hands, did a teacher's course and obtained employment in a technical college in Leigh. He cycled ten miles each way to and from work every day. Anyone getting a job was the talk of the village. John Ball, the grandfather of my flying friend Francis, cycled an incredible distance to work, spent his day laboring in a brick works and then rode home each day. Given the circumstances of most people in our village, and the economic times in general, I was very lucky indeed, and I appreciated the opportunity to be paid to learn.

I enjoyed my time at 'the Tech'. The work required a lot of effort, but the fact that I liked what I was doing made it all worthwhile. My time was not all taken up with study, of course, and I had many other interests. My allowance was half a crown a week, which covered my bus fares of three pence each way for five days; but if I went on my bike, I would keep getting fit for a dream I had not forgotten, and best of all I could save the money for more interesting things, such as half a pint at the Red Lion hotel, or a trip to the movie pictures.

Foremost amongst the list of 'interesting things' were girls! Unfortunately they tended to cost money and time, both of which were in very short supply. However the competition all had the same problem. I had known many of the girls in the village as playmates from earlier days but they seemed to have changed into quite different creatures like butterflies—not that any of them had started as hairy little caterpillars—and they were now all very beautiful in many different ways. The girl whom I most admired was Lilly. I had admired her from my first day at school. Alas she was still totally indifferent to my advances.

Then there was Marion Holt, who lived close by and who had always been friendly. Marion was a playmate with my cousin, so I saw quite a bit of her and she and I would often go for a walk through Lord Crawford's estate on a Sunday night after she had returned from the evening service in the Methodist Chapel. Of course in a small village one of the most popular occupations is gossip. Marion was not immune to that and so she was called to account by two of her aunts who were of the High Moral Chapel congregation. Highly moral they may have been, but they were not beyond listening to rumors. Marion was not amused by the assumption of pregnancy, which in fact was totally unfounded. I have no idea how or why the rumour even started; and she really was upset by all this. Marion and I extended our walks to the more visible areas of the village, but I do not know if the rift in her family was ever healed. There was also a very good-looking young lady, Janey, who was regrettably, far too serious to be taken seriously. So I pursued my friendship with Marion and kept my eye on Lilly, but all the while, what I really dreamed of was flying.

Chapter Eight
Preparing for War

ON 3RD SEPTEMBER, 1939, THE war which we had been anticipating, was officially announced. Mr. Chamberlain's 1938 visit to see Hitler had not fooled many people and 'peace in our time' was colloquially translated as 'peace for a short time'. We all knew the second World War was coming, it was just a matter of when it would officially start, so although I was busier than ever, helping mother with the house and studying a heavy engineering degree, I made reading Hitler's book a priority during 1938. The book, *Mein Kampf*, or *My Struggle*, chronicled Adolf Hitler's life alongside excerpts of his speeches and an exposition on his political ideology—that which he had made known.

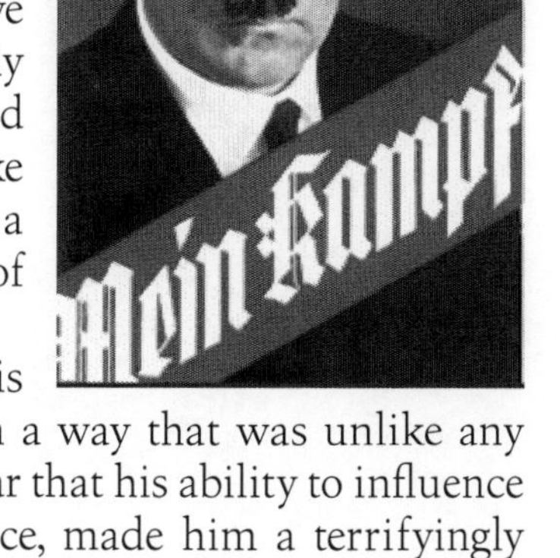

I had heard that Adolph Hitler was a very charismatic and persuasive man, and this book certainly confirmed such was the case, and the extent of his ability to make convincing arguments was without a doubt the most frightening aspect of what I read.

Hitler had conveyed his messages, throughout his career, in a way that was unlike any other political leader, and it was clear that his ability to influence people, and win their full allegiance, made him a terrifyingly powerful man. This gave me some insight into the mentality of Germany and, in particular, the thinking of her civilians and the lower ranking men and women who were sent out to do most of the actual fighting.

After losing the First World War, Germany had been brought to her knees. Through reading Hitler's book, I understood why there was so much anti-Semitism. Many Jewish families from Poland and the surrounding Eastern nations had migrated to Germany in the years that followed the war, and because the Jews formed a community of their own and didn't intermingle a great deal with the Germans, they were seen as separate and became a very easy target for resentment. The German civilians had been left desperately impoverished by the war and saw the Jewish people's business success and were upset about the migrants taking over many of the local businesses and employment opportunities. The new Jewish owned businesses were seen as taking people's jobs away, and in the context of such desperate economic times, this was a significantly emotive issue. At first there was a relatively low level of resentment, but under the leadership of a man who had his own agenda and who was able to manipulate situations to his advantage, it wasn't long until animosity was fanned into flame and then it became something entirely other than what it had started as.

There was no simple, single explanation for the situation that was simmering away and of course it depended on who you were and where you came from, as to how you would view the situation. When I tried to look at it from the perspective of the German citizens, it was clear that after they had lost the First World War, the Germans were given reparations that they had to pay, but which they had no means of ever paying, and so they ended up saying, 'we are not going to pay!' Perhaps worse than this, the heavy burden of repayments created an environment which fuelled resentment and racism and made them susceptible to a dictator, who presented as though he had the answers. To me the whole situation was predictable, given the way things were left after the First World War. It was as though a playwright had set the stage for a sequel. You did not need to be a genius to predict that it would happen.

There was no doubt about it; Adolph Hitler presented the Germans with what looked like a solution. In the beginning, from the perspective of many of Germany's citizens, things worked under his leadership. He took a ruined country and within a very short period he got the nation back up onto its feet. They would never have been able to start a war if he hadn't. He was able to make the circumstances of the time work to his advantage, appearing as though he stood for 'discipline' and 'employment opportunities', he skillfully provided economic scapegoats and leveraged, to his advantage, the people's fear of a possible Communist takeover. It didn't take much to unite and motivate the Germans to embrace his leadership. Karl Marx and Mussolini did the same; they were in desperate situations and presented the people with both scapegoats and answers they could believe, and because people at that time didn't have the advantage of knowing the history that would unfold, or much access to other information, they had no reason to question their leaders.

Furthermore, Hitler certainly wasn't sharing his full intentions, but rather, only that which he thought the people were ready to hear; that which he had prepared the people to hear. On the face of it, this charismatic leader was offering a way forward, through 'discipline', a value that was embedded in the German psyche and the 'creation of employment' something which they desperately sought, so I could understand why so many people would have been drawn to his leadership.

With this insight I could see that the German people were ordinary men and women, many of whom were hard working, family orientated citizens whose culture and values were not all that different to ours. From what I read, it was disturbing how easily most of the people had been manipulated. It wasn't as though all Germans were fundamentally evil, or even naïve and unquestioning. They were more similar to us than anybody dare suggest. It was only that they were in very different circumstances and oppressed under a heinous leader, a leader who presented as a man who provided solutions, but who really understood the power of mass psychology and how to gain people's allegiance by only giving them small pieces of information. There was a reason why so many tens of millions of men and women were manipulated into thinking the way they did, and I realised that if it could so easily happen to so many, then it could have happened to any of us, if we had faced the same conditions.

Amongst my peers, friends or family, I was the only one who chose to read *Mein Kampf*, and at first I encouraged them to do the same, because I thought that it was very important that we to try to understand the underlying reasons for the war. Yet, as I drew towards the end of the book, and the start of the war felt increasingly near, I realised that reading Hitler's book had done me the greatest disservice; now I could understand why the Germans were acting the way they were. I did not believe that they were a population of nearly seventy million evil people. Rather, there were only a few, but they had managed to completely deceive and exploit the masses.

With the Second World War increasingly imminent, this realization put me in a hard position. I kept thrashing it over and over in my mind, searching for the right thing to do. On the one hand I could see that what the German people were doing was truly heinous, but in most cases they were ordinary people, foot soldiers, who had been terribly manipulated, and they could not speak out without fear of serious retribution. Many of the intellectuals and thinkers who had gone before them had tried, and they had been promptly 'gotten rid of'. When I read excerpts from Hitler's speeches, I could see that he had been addressing normal, family men and women, but he knew how to stir them up. He understood the power of fear, and he knew how to work on them to gain the allegiance of the majority or at the very least, their silence.

I struggled to imagine myself fighting them on a personal level. I didn't want to kill anybody, much less now that I had an understanding of why they were acting in this way. It was truly awful and so difficult to know what to do. How could I justify killing other sons and daughters, mothers and fathers and even children when I couldn't honestly answer two fundamental questions, 'Would I have been any different?' and 'Would I have stood up and said "No", in that context?' I tried to imagine being fed the propaganda that they were, and being under the tyranny of such a leader; it was hard to believe that I would have acted any differently.

So I asked myself again and again, how could I justify killing them and would I really be able to do it?

Before we were officially at war, it was difficult to imagine fighting and killing the German people and even harder to know what was the right thing to do. I knew that things are never entirely what they seem and a politician always paints the sort of picture they want to, but as the war was approaching I had to take a stance. After I read more articles and confirmed their claims with some refugees I knew from Germany and Spain, I came to the conclusion that even if you felt empathy for the Germans, the fact that Hitler's regime was so brutal in its treatment of the Jews demanded action. His aggression and brutality was what made us feel obliged to stop him, and the only way we could see to stop Hitler was to stop the German people. It was an awful, confronting reality.

With what had already started in Europe, even before the war was officially announced, we knew there would be no limits of the brutality that the Germans were capable of, even if they were just foot soldiers, or young pilots following orders. Despite this it was still very difficult for me to come to terms with killing anyone. As I understood it, our job was to kill them before they killed us, but it wasn't as simple as that. Many of them were probably nice, young people, who also didn't want to go to war. It was a very difficult process to work through and try to come to make sense of. It seemed to me that other than a very small number of people with a vested interest in the power and profits that come from war, nearly all of us were soon to be fighting a war we didn't want, and of those who stood to profit from it, I didn't see too many of them out front and centre.

Across our nation, before the war, there had been a bit of sympathy for the Germans, with a lot of British people agreeing that after the First World War and the Treaty of Versailles, the Germans had been given a raw deal. However, it wasn't long until the citizens of the UK had seen enough to feel convicted about their decisions to fight. There was no peer pressure to sign up on our side, but most people did see the merit in fighting; we knew what was going to happen if we didn't. Knowing its inevitability two years before the official declaration, by 3rd September 1939 I had been slowly prepared for the madness that is war, and I had made up my mind to enlist.

Hitler's book had provided an insight into psychology and power that I would never forget and a framework for understanding much of what would unfold, but for now all I knew was that something very dark and heavy was brewing. I could not yet imagine how it would play out, or on what scale, but what I did know was that fear among the English people had reached almost tangible levels. The First World War was still fresh in the minds of many people and for those of us who had not lived through a war, the newspaper and wireless reports of current events in Spain and Poland had proved more than enough to send shivers down our spines. Poison gas had been used during the First World War, but by now the Germans had even more advanced techniques, and based on what had been happening in the Spanish Civil War over the past three years, there was a great fear that poison gas and air raids would be our demise.

In response to this, in 1938, everyone was issued with a cardboard box, just larger than a tissue box. Inside the box was a full face respirator. We were not yet at war, but receiving these awful looking gas masks, sure made it feel very imminent and real. The brown facemask covered the front and top of the head, and protected the eyes through its inbuilt goggles. At the front of the mask was a single filter that made the wearer look alien and the large goggle eye shield did not help the eerie appearance. We were grateful to have been issued with these, but they did nothing to quell public fears. People were now talking about the coming war everywhere and you could hardly have a conversation with anyone in the local village without the subject being raised.

We were told never to go anywhere without our respirators, and we didn't. At some schools and workplaces they had mask drills where students had to put them on immediately and practice breathing through it without fogging up the goggles, but we did not have these at 'the Tech'. At some places they even recommended wearing it for fifteen minutes a week, just to get used to functioning with it on.

By the beginning of 1939, the government began to issue

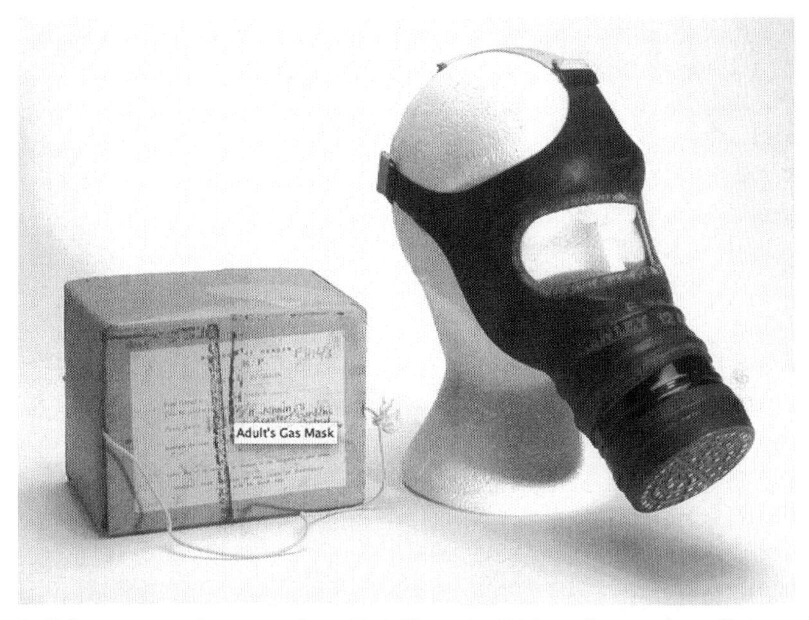

Full face gas mask—issued to all civilians in 1938, and carried at all times slung from the shoulder in a stout cardboard box.

Below: War poster reminding people to always be prepared and carry their gas masks with them everywhere they went.

millions of Anderson Shelters. Families whose income was as low as ours were issued these shelters for free, whilst others had to rent them from the government. These galvanised corrugated panel kits were designed to protect and accommodate up to six people in an air raid. The instructions in the kit told us to dig a large hole about 4 feet 6 inches wide by about 6 feet 6 inches long, with the hole itself being at least 4 feet deep. We were to erect these kit shelters in the holes. Then around the three sides that did not have a door and

over the roof we would pile up a minimum of 15 inches of soil. The Anderson shelters were another sign of something truly ominous, but surprisingly in our village, people reacted better to these than they had to the gas masks and some chose to make the most of it by decorating their dirt mounds with flowers and fresh plants. In some areas it even became a friendly neighborhood competition. The issuing of these bomb shelters and the gas masks was strange, because although I would consider myself rather pragmatic, and would be the first to accuse people of over reacting, at the time such protections actually felt necessary. We had seen what Germany was capable of, and we knew she was about to attack.

In the few years leading up to the beginning of the war, England had seen the construction of many 'shadow factories' for the production of war material. These factories were scattered throughout the country to avoid providing single targets, which if struck, would significantly impact the supply of war material. Rolls Royce had one such factory at Crewe, Cheshire, about 40 miles from Wigan. Once the war had officially started, they increased production and it was here that my brother John found employment, inspecting the construction of Vulture engines. Around the same time my sister Elizabeth also went to work for a De Haviland's factory, making propellers.

There was an enormous national focus on manufacturing tools, weapons and vehicles for the war. The shift in labor efforts from farms and food factories to military activities, in addition to the reduction in Britain's imports, resulted in food and clothing shortages and hence rationing, and of course this was worsened once we were officially at war. Eventually the German submarines would specifically target some of the supply ships bound for England. In fact, some records suggest that prior to the war, England had imported 55 million tons of food per year, but one month after the war started, imports were decreased to an average of only 12 million tons per year. The government keenly promoted turning ornamental gardens and public parks into productive plots, and so the term was coined, 'Dig for Victory'.

The government then issued everyone with a little ration book and when we would buy particular items the shopkeeper

Promotional material encouraging the growing of food on every available plot.

would cancel the corresponding coupon or tear out that page. Two months after the war began, petrol was rationed. January 1940 saw the rationing of butter, bacon and sugar. As my contribution to the war effort I voluntarily ceased having sugar in my tea, but it didn't help much, because by July 1940 even tea was rationed. I have never taken sugar in my tea since. By March 1940, all meat was rationed. As the war progressed the same happened with eggs, cheese, lard, cereal, rice, and fruit as well as soap, clothing and coal. The rations were enough, but only just.

My siblings and I could see that a war would be tough on my mother and that rationing was likely, so for a long time

Cyril's Mother, Aida Cooke, tending the garden.

before the war had officially been declared, we had started stockpiling flour, sugar, butter and anything we could not grow or make ourselves. This gave mother a bit of a head start, but it was never going to be enough to see her and Emma through the rationing periods. This time was particularly hard on my mother who was still grieving the loss of my father, and having already lived through one World War, she knew more than we did about the trials we would soon face. Fortunately she was largely

Cyril Johnson fixing the furniture with a hand-powered drill and plane.

self-sufficient and between the abundant fruit and vegetable garden, the chickens and the bee hives, they had enough food, and mother didn't mind tending the garden. John, Elizabeth and I spent a lot of time tending to maintenance issues around the house, so that mother would be okay if and when we were called away for long periods at a time.

Cyril Johnson, with the family dog, Winki.

With so many men leaving their farms for military duties, the Women's Land Army (WLA), was restructured, after it was initially formed during World War I. During World War II, the WLA trained and equipped more than 90,000 women to work on British farms. They were able to do all the usual farming activities, which were typically seen as 'men's work'. They were trained in threshing, plowing, sowing, harvesting, driving tractors and all manner of other activities needed for food production, and as a result, they were able to grow much of the nation's food, without which everything would have come to a halt. It was strange how important the simple things in life had become, and the basic things that we could not take for granted. Everything had to be organised and prioritised and considered on a large scale.

By the beginning of the war we already had many refugees in England, one of whom was a man named Kurt who had become a new family friend. He was Austrian and he had been in charge of the guard in the hotel where Adolph Hitler stayed in Vienna. Kurt was well respected and a part of a very select circle of military personnel. Suddenly, someone discovered that Kurt's mother was part Jewish. Overnight his luck had changed. He was now a legitimate target. Fortunately he had friends in high places and they gave him a map and a compass and said 'Switzerland is that way, good luck'. When he eventually made his way to England, he somehow came to settle in our remote village, and once he discovered that my mother, although English, spoke German, he often came to visit and just enjoy a conversation in his native tongue. He had some interesting political perspectives, but most notably he warned my mother never to underestimate the Germans, saying, 'they won't hesitate to torpedo a ship full of evacuated children', which is in fact what did happen in 1940.

The whole of the United Kingdom seemed to be preparing for the worst. Soon many of the children were evacuated out of London. This must have been a huge ordeal for them, because they were frightening and uncertain times for an adult, never mind a child being sent off to live with a new family. Mother had listed our house on the potential billeting list, not that I can imagine Emma would have been excited about the idea. As it turned out, there were already more than enough other families in Wigan who had volunteered. Suddenly loads of children arrived, with the hope that they would be safer in the smaller towns, where the chances of an air raid were much lower. London was an obvious target and received her share of bombing.

Wigan was not a likely target, although nearby Liverpool certainly was and from time to time we received the odd stray bomb, as some of the German bombers got lost. The first bomb that was dropped over our village landed in a field and killed a cow. The second was dropped near the steel works, which may seem like a logical target, except that it had not been operating for decades. Nonetheless, a bomb actually landed out the front of the pub and blew the front doors off. The patrons were all very shaken up, but there weren't any significant injuries.

My sister Elizabeth was in the Women's Volunteer Service (WVS) at the time so she was called to assist these patrons. Being home and otherwise unoccupied that particular day, I was roped into it too. After that effort, the WVS decided that I was pretty handy, seeing that I had a car, a beautiful Morris, and a bit of spare time, but most importantly, access to some petrol. The supply of petrol for civilians had stopped almost immediately after the war had started, but my brother John had been stockpiling it in grandfather's garage for months, knowing that this would likely happen. Furthermore, drawing largely from

John's expertise in working in the Rolls Royce engine factory and my understanding of hydraulics and pressures, between the two of us we worked out that the engine would be more fuel efficient with a water injection system. It was a fiddly business, but we put a tank of water in the engine and bypassed it through the exhaust manifold such that it became steam and then emptied back into the carburettor. The result of this was petrol and steam, with just enough water to make the mixture of petrol and air wet. This meant that we would get a better expansion ratio when it would burn and the petrol would ignite, creating more pressure, and therefore making it more efficient. We put a lot of time and effort into this and eventually it paid off, because it helped reduce the amount of petrol we used, and so it was that for the first few weeks of the war I became the driver for the WVS while I sorted out my own applications and tried to find a more appropriate role in the war.

My dream of being a pilot had never been far from my mind, but this time there was a new stumbling block that frustrated my attempts to join the Royal Air Force. When I tried to join, I now had first to be passed by the representative of the Minister of Labor and because I was an engineering student I was not allowed to get as far as the recruiting officer's door. They already had a defined pathway for engineers, and I was advised to join the Royal Engineers. Apparently that branch of the army was looking for people with some engineering qualifications to train as officers. I had no ambition to join the army but thought that this could be better than waiting to be conscripted into some ghastly muddy trench. My application was accepted by the army, with more enthusiasm than I had for their corresponding offer; but even this was not without its strings; I was to attend another medical.

The ancient medico detected the leaking valve and decreed I had to be referred to a specialist. He escorted me to the door and putting his arm on my shoulder, told me 'not to worry as you *might* live to be an old man of seventy!' I continued my fitness campaign with even more vigor and now played rugby on Wednesday and Saturday afternoons plus training, which I thought would either kill or cure me for certain. I eventually saw the specialist that the army had sent me to and he thought I would be acceptable. Now that really created a problem: I did not want to join the army!

One day Aunt Teresa, my mother's sister, was having lunch with us and heard me speak of the injustice of a system which seemed determined to send me where I did not wish to go. It seemed wrong that the RAF did not want me but the Army did. She asked what the problem was and then said she would have a word with a friend of hers. The friend was the local Squadron Leader in the recruiting office. I duly got the message to present myself at a certain time and I presume with some help from persons unknown, though 'unfit for duty in every flying category', I had evaded the Ministry of Labor and was accepted into the RAF.

I was told to proceed the next week to Padgate. Here I joined a large group of hopefuls all wanting to fly aircraft for His Majesty, King Edward VIII. This time I knew what to expect from the medical examiners and my heart behaved itself due no doubt to the threats I had made to it, well, that and thanks to whoever was pulling the strings, that's what my record showed! What did surprise me however was the number of very clever and fit looking characters who were not accepted; by the end of the day there were not many of the original group left to take the oath of allegiance, but I was in!

The entrance to Padgate—Cyril's first view of life with the RAF.

CHAPTER NINE
Initial Training Wing

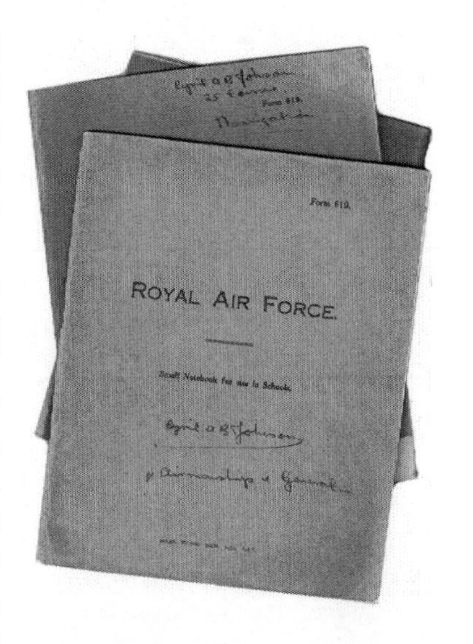

AS NEW RECRUITS OUR WELCOME to the Royal Air Force or 'the RAF', was a four week 'basic training' course. Our days started with a Non Commissioned Officer (N.C.O.) entering and walking the length of the hut and shouting 'Wakey, wakey, rise and shine!' That was followed by a trip to an adjacent hut, 'the ablutions', where one shaved and washed in icy cold water. Then, once we were properly dressed, we were ordered to go for breakfast, which was quite adequate, before returning to make our bed, clean and polish the bed space and set out our kit and 'biscuits'—a mattress that came in separate pieces to be folded and transported when necessary. It was quite a routine, but we soon got used to it. Everything had to be exactly in its correct place, cleaned and polished, and I mean exactly!

After everything was clean and tidy the 'on parade' call would come. They taught us 'square bashing' which consisted of marching, halting, about turns, and dozens more things which would become second nature. Physical Training (P.T.) was an integral part of the training, as were route marches and bayonet practice. The latter was something that amazed me; I thought I had avoided all that, but not so. The RAF was formed from the

British servicemen undertaking bayonet training, c.1940.

army and many of the army's methods and traditions were still in effect. To that end, we were escorted to where they had several large bags of straw dangling from a framework. These were to be the 'enemy' and we were instructed to charge at them screaming at the top of our voice (yes, that was the instruction) and stick the bayonet into the enemy. It worked ok, charging at an inanimate object, but I did wonder what the enemy would have been doing in response if they had not been bags of straw.

Route marches might have been fun in summer time, but in winter they were bitterly cold and windy. The winter of 1939 was particularly bleak, and only a short distance away, in our home in Wigan, it was cold enough that mother had been snowed in, with several feet of snow. The town's council workers duly arrived, but the snow was so deep that even the plough got stuck! Fortunately my brother John was available to assist the plowmen, so I didn't feel too bad about not being home to help. It was one of the worst winters we had endured for many years and 12 miles away at Padgate, the temperature seemed no better. Despite this, P.T. was for all weather. Wars didn't stop on account

The family's driveway on Haig Road, Aspull with John Johnson assisting to free the overwhelmed snow plough.

of the cold, so I suppose it made sense that they would not relent on our outdoor training either. It was a struggle to dress with enough layers to take the edge off the air, which was cold enough to make our fingers and cheeks sting, while anticipating the heat and sweat that would inevitably come from running several miles.

One march was an event to remember. After a few miles we came to an open-air swimming pool. Naturally, with the winter weather, there were no customers, none that is, except us. We were marched through the gate and lined up along the side of the pool. How it hadn't frozen I am not sure, other than to suggest there must have been an awful lot of salt in it. We had difficulty believing the next command until it was repeated with the appropriate expletives. The order was to strip—naked—and swim across to the other side. Those along the shallow side were not as lucky as they thought being told to 'swim, not bloody walk'. We were then allowed to get dressed which was another difficult operation with such shaking limbs and damp skin; and the return to camp was done as a run, so this kept us warm at least. I now knew why the Air Force required people in perfect health; anything less and we would not have survived the initial training.

When we first arrived we had been issued with our kits but of course with everything being on such tight rations, it wasn't long before other people's things started to 'go missing'. One morning I found my knife, fork, spoon and mug were not where I had left them. Our N.C.O. seeing me empty handed, looked pityingly at me and said, 'Well you know what to do'. 'What?' I asked. 'Well, someone pinched yours, didn't they?' he replied, thinking I must be pretty dumb. 'Oh! I see' I responded. So that night when he went out, I entered his room and pinched his! He never said anything but I suspect he would have liked to kill me.

The four weeks ended with a passing out parade. It was complete with a physical training demonstration, and a fife band, which made use of a small high-pitched flute-like instrument similar to a piccolo. We had learned a lot in four weeks, but perhaps one of the most useful things we had come to learn was how the ranking system worked. At that time, those of us who joined the Royal Air Force were usually joining the Royal Air Force Volunteer Reserves, distinguishable by the VR badges stitched or pinned on our uniforms. Being a volunteer meant that your ranking was determined according to the Non Commissioned Officer (N.C.O.) system, in which ranks ascended as follows: Leading Aircraftman, Senior Aircraftman, Senior Aircraftman (Technician), Junior Technician, Corporal, Sergeant, Chief Technician, Flight Sergeant and Warrant Officer. People did not always 'climb the ranks' in this particular order, because it depended on the area they were going to specialise in.

Meanwhile, Commissioned Officers, who were professionally employed and therefore not part of the Volunteer Reserves, had a separate ranking system, which went as follows: Pilot Officer, Flying Officer, Flight Lieutenant, Squadron Leader, Wing Commander, Group Captain, Air Commodore, Air Vice-Marshal, Air Marshal, Air Chief Marshal, Marshal of the Royal Air Force. In some cases, the Commissioned Officers out ranked the Non Commissioned Officers, although this was not always the case. Many of the higher ranks within the Non Commissioned Flight (N.C.O.) system actually had more authority than those of the Commissioned Officers. For example, an N.C.O. Sergeant would have a greater level of authority than a Commissioned Pilot Officer, but only sometimes and it depended on the situation. It was a bit of a confusing system, but we soon learnt it, and if we made a mistake, we were quickly informed. At this stage I was at the bottom of the heap, under either system, but I knew that in due time, when I would eventually climb the ranks, I would need to know my place or accidently disrespect someone, and the outcome of this was likely to cost more than the time and effort it took to learn the system in the first place.

To make things even more complicated, many Non Commissioned Officers became Commissioned Officers. For example, if you were going to be a pilot, you started as an N.C.O. Leading Aircraftman until you finished your training. Then you would be given your 'wings', a badge which pilots would proudly stitch onto their uniforms. Following this, one either became a Sergeant Pilot, who was still a Non Commissioned Officer, or a Pilot Officer who was a Commissioned Officer, but how they determined which one we were to become I am not entirely sure. I suppose that they assessed suitability for Commissioned Officers based on education and leadership qualities and behavioral reports, but what other factors came into it I have no idea, because if it was purely based on behavior I don't think I would have gotten particularly far. In any event, I became a Commissioned Pilot Officer soon after I finished my pilot training, but at this point I was still a Leading Aircraftman.

Perhaps due to the winter weather, the Air Force was unable to send us to the next place for our training and so we were made temporary clerks, in my case dealing with records. This revealed some interesting information, for example our Sergeant had money stopped from his pay to support some girl he had 'gotten into trouble'.

One morning, after a couple of weeks, we were given a bag of 'rations', mostly sandwiches, and put on a train. We were not told much about what to expect, although I wasn't particularly worried, because I knew I still had a lot more training in front of me before I would be sent out to war.

Late that night we arrived at Babbacombe, a small seaside resort near Torquay on the South Coast of England. I proudly filled in a section of my log book, under the heading Record of Service, that I was now officially on the number 8 Initial Training Wings course, Torquay, and the starting date, 23rd November 1940. We were billeted in what was previously a private hotel. Here we had an enjoyable few days taken up with lectures on health, which mostly covered venereal disease and how not to get it, with the lecturer commenting that 'our girls are safe'! Drill and P.T. occupied a little time, as did taking of blood samples and giving of inoculations, for common diseases like diphtheria and small pox, but other than that we had a lot of free time. On the last day they taught the administration of First Aid in the event of an injury in an aircraft. I thought this was a little optimistic considering the dangers we were soon to face and the lunchbox sized first aid kits we had to play with.

After a couple of weeks we found ourselves on a train again, this time bound for the Initial Training Wing, at Newquay, in the South West. Once more we were billeted in private hotels, which lined the front of this erstwhile holiday resort. I shared a room in the Terrace Hotel with four other aircrew cadets. They were: Arthur Boddam-Wetham, whose family all seemed to be Admirals and Air Commodores; a red-haired Cornish man who, not surprisingly, was called Ginger Jones; an Irish lad named D. J. Hinds and Dennis Clarke who was a very precise sort of man. It was still the depths of winter but the physical training we did, often on the beach in the wind and rain, must have done us good because none of us fell ill with common colds or 'the flu' despite living together so closely.

When my roommate, Dennis, had been an undergraduate at Cambridge he had learned to fly a helicopter, an auto-giro. At first it came out as something he was rather proud of but then part way through his story he thought better of it, and we were sworn to secrecy. He feared, with good reason, that if the Air Force ever found out, he would be excluded from the training on fixed wing machines, and sent on to fly a helicopter—which in

Left to Right: Arthur Boddam-Wetham, Lance Holliday, Ginger Jones, D.J. Hinds, Dennis Clarke, and Cyril Johnson.

a war wasn't an attractive option. Arthur was a very kind man who had a natural teaching gift, so he was always helping other students, and I was grateful to have him as a room mate. Standing with us in the photo was Lance Holliday, whom we called Uncle Rector because he always knew everything about everybody and what was going on, a very useful friend, considering the lack of communication that takes place on these training courses.

Also on our course were two pilots with whom I would become very close pals. Firstly, I met a young man from a town near Essex called 'Clacton-on-Sea' whose name was Maurice Carter. Maurice soon became Maggie as in Magna Carter, and from then on he was always known as Maggie. Then Ron Bernard introduced himself to me, he was a very determined man who was quite likable. Ron actually struggled a bit, and they almost dropped him from the course on a few occasions, but he persevered, and studied harder than any of us, so in the end, he graduated as one of the top men in the course. Maggie, Ron and I did all our training together and were eventually sent overseas in the same unit, so we become tight mates.

Group 8. Initial Training Wing, Newquay. (Cyril Johnson, 3rd row, 9th position from left, with friends Ron Bernard, 1st row, 8th position from left and Maggie Carter, 2nd row, furthest right).

The most imaginative man on our course was an ex-naval officer who managed to think up the most incredible stunts to tease the authorities. Why 'the ex' we had no idea until much later when an explanation was provided by someone in the know. The story was that when he was a Naval Officer, whilst on watch on the bridge, he would use the communication system to play battleships with the Engineer Officer in the engine room, an arrangement which was certainly not known to his superiors. Then one night, while battle games raged via the speaking tube, one of His Majesty's destroyers ran aground! Presumably the Air Force were not concerned about their aircraft running aground but they may have also overlooked the effects of one so ingenious on the 'good order and discipline' of their service.

Being winter and with a black-out enforced, our first parade was held in front of the Trenance Hotel in complete darkness. 'The ex' had acquired a mechanical shaver, similar to the modern electric shaver but it operated by squeezing a lever on the side. He was always running a bit late for things, so he went out on parade still shaving. Apart from the noise of Corporal Jones trying to put the rabble before him into some sort of order there was the high-pitched mechanical noise from the shaver. Jones asked, many times, but never found out where the noise was coming from. I don't imagine he would have been happy if he had.

The 'ex' was wonderful for our sense of mateship and belonging. Coming from a position of authority in the Navy he knew what it was that would annoy our leaders and how to execute it perfectly so that no one would get into any real trouble, either because they couldn't find the culprit or more often because it was quirky enough that they couldn't really justify punishing the lot of us. One day while we were being marched in well-ordered and straight lines down the main street, 'the ex' arranged for us to all simultaneously start to limp on our left foot. It must have looked hilarious as we all did it in unison, and our timing was perfect.

We quickly developed a routine and each morning after our parades we were marched to our classes or drill or got changed into our P.T. kits. Our lectures comprised of Navigation, Airmanship, Signals and a variety of other things the Air Force thought would be good for us. At this point our aviation training was all classroom based. I filled up several exercise books worth of notes and I must have been rather keen because in my books I had drawn diagrams of how all the controls relate to the surfaces of the aircraft, how an internal combustion engine works, and filled up page upon page about all the subjects we were taught, including swinging a compass, cross-country flying, and plotting position lines by direction finding (D.F.). The aerodynamic theories were all new to me, but after I thought about them for a bit, they were mostly just a matter of simple physics and they made a lot of sense. The concepts were not too difficult to grasp which was a good thing because it wouldn't be long until I would be up there on my own and unlike my early school days, I now had a serious motivation to learn.

Almost seventy years after the Second World War, it is surprising to see that much of what remains a core part of a private pilot's theory lessons has not changed. The principles are essentially all the same, except that during the war, there was a greater emphasis placed on certain subjects and a few additional things that we had to be aware of. Foremost on the

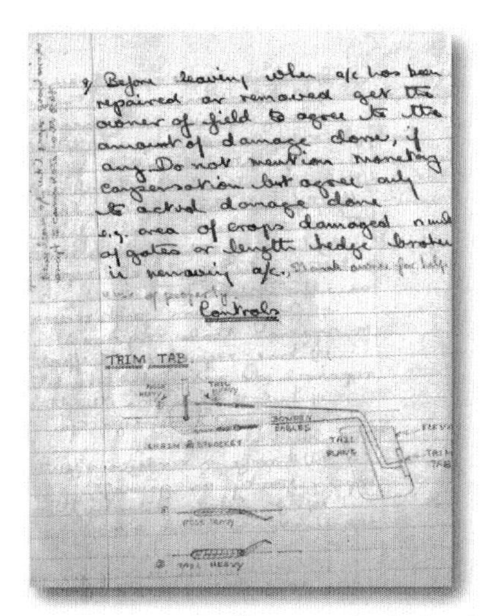

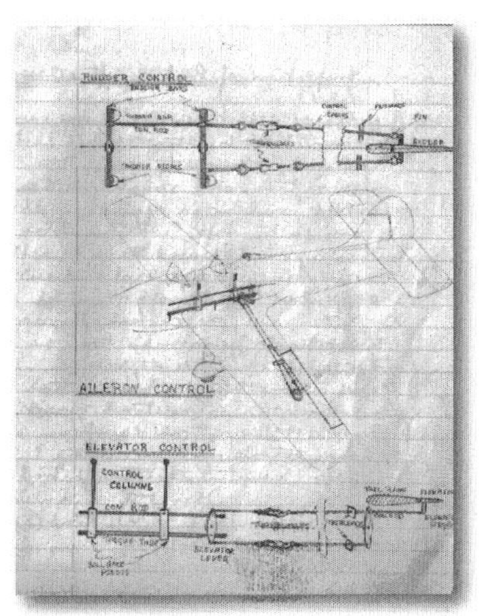

Cyril's notes depicting how the joystick is linked to the control of the surfaces of the aircraft.

list of additional lessons was the identification of friend or foe aircraft (IFF). My 1940s 'state of the art' IFF system, comprised of a long list of visually distinguishing features of the enemy and allied aircraft. There were quite a variety of aspects and all needed to be committed to memory for easy recall at a moments notice. I diligently made note of which aircraft had its engines mounted and where in relation to the wing struts, the relative thickness and angles of the various struts, and the number of engines, rudders and trim tabs. I even noted that 'all German 4 engine bombers are low wing' aircraft, although this couldn't be solely relied upon because so were many of our allied bombers.

Then we were taught what to do in the event that enemy action should cause the runway lights to be extinguished, and told to simply circle around the nearest beacon at the correct height, until it was all clear again. The instructor mentioned, calmly, that while we were circling, if we happened to be picked up by the airfield searchlights, we should fire off the colours of the day, for identification purposes, and our own men wouldn't shoot us down. With any luck, when it was clear again, they may even dip their lights in the direction of the aerodrome to guide us back in. It all sounded a bit surreal.

RAF student pilots learn how the engine works on a mechanical level.

Amongst those things considered as a normal part of a private pilot's license, but which were perhaps emphasised to a greater extent, were, fire in the aircraft, forced landings when over unknown territory and indecipherable communication. Having learnt of all the ways in which the enemy can shoot down a

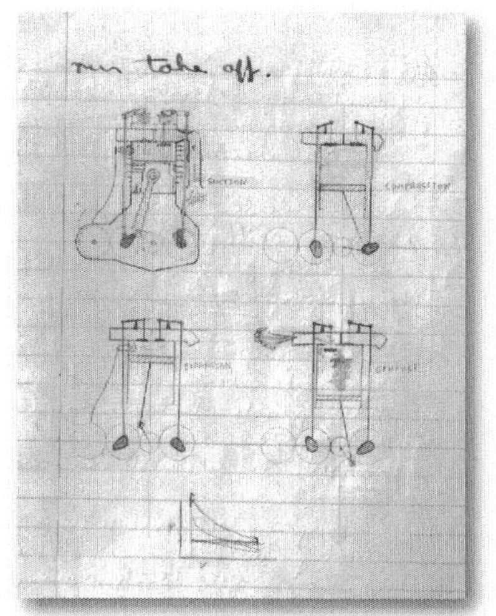

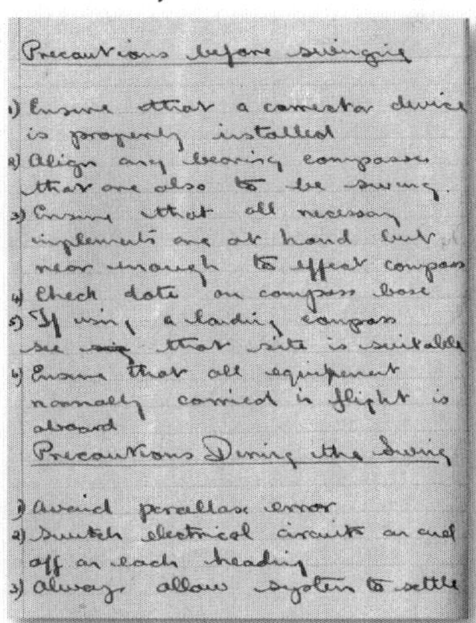

Precautions before swinging

1) Ensure that a corrector device is properly installed
2) Align any bearing compasses that are also to be swung.
3) Ensure that all necessary implements are at hand but near enough to effect compass
4) Check date on compass base
5) If using a landing compass see that site is suitable
6) Ensure that all equipment normally carried in flight is aboard

Precautions During the Swing

1) Avoid parallax error
2) Switch electrical circuits on and off on each heading
3) Always allow system to settle

Left: Cyril's notes depicting the workings of an internal combustion engine.
Right: Precautions before and while swinging the compass.

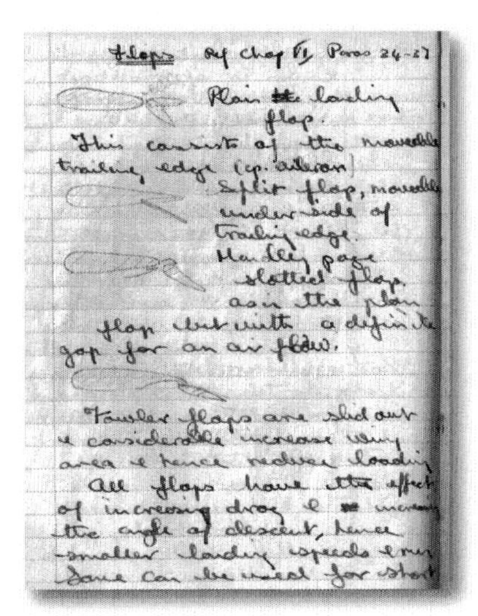

Flaps Ref Chap VI Paras 24-27

Plain landing flap.
This consists of the moveable trailing edge (cp. aileron)

Split flap, moveable under-side of trailing edge.

Handley page slotted flap. as in the plain flap but with a definite gap for an air flow.

Fowler flaps are slid out & considerable increase wing area & hence reduce loading

All flaps have the effect of increasing drag & increasing the angle of descent, hence smaller landing speeds & ... Some can be used for short

The difference between Fowler and Henley-Page flaps.

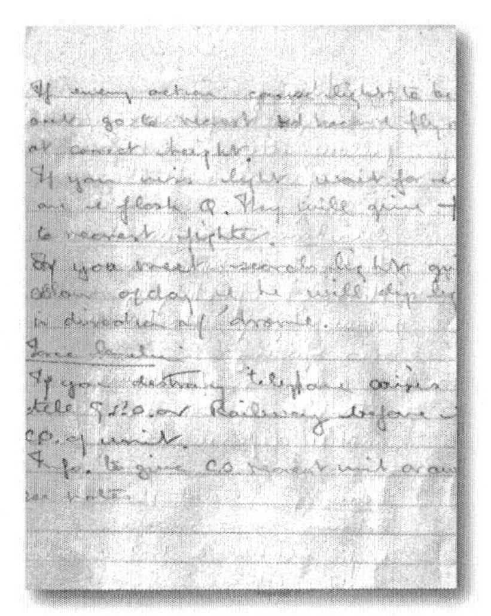

Instructions in the event that runway lights are extinguished due to enemy action.

plane, from the ground or the air, I can't say anything pleasant sprang to mind when I wrote the following regarding fire; 'Open throttle wide to assist in using up petrol... after engine has ceased

Engine Starting & Airscrew Swinging
Hand swinging & impulse starter
Chap II paras 41 & 43.

Action in the event of fire in the air

Chap. III Paras 248.-251.
a) Turn off petrol
b) Open throttle wide to assist in using up petrol in carburettor as soon as possible
c) After engine has ceased firing switch off ignition
d) Sideslip to keep flames away from fuselage. Use extinguisher when possible
e) look for a forced landing field fly towards leeward side
f) If the fire gets out of control & the height is sufficient abandon aircraft by parachute endeavouring to set controls so that aircraft will maintain a steady glide towards open

Action in the event of a fire in the air.

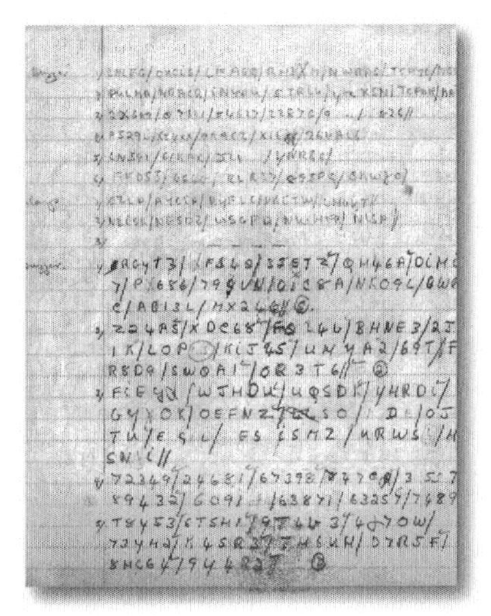

Cyril's practice exercises interpreting Morse Code from an Aldis lamp.

firing switch off engine... sideslip to keep flames away from fuselage. Use extinguisher if possible. Look for a forced landing field'. Finding a landing field wasn't without its problems either, depending on whose territory you were over, and who saw you. You may or may not wish to be found. Then of course it may be necessary to communicate in the dreaded Morse code, which was designed to be confusing, but as a dyslexic, I would create my own uninterpretable code.

Flying School had an ere of seriousness, but of course, we always managed to have a bit of fun, and often joked about the madness of the whole situation. Dennis and I were supposed to be 'the better educated' with regards to mathematics, having had some university standards of achievement in that arena, so it was a surprise to the instructors when a test put us both at the bottom of the class. The explanation caused no little mirth; neither of us could understand the meaning of such obviously simple questions!

Apart from aviation, there were a lot of other things that we needed to learn in order to get on okay once we were out there. Our Armaments Instructor was an N.C.O. who in his former life had played cricket for England. The Officer in charge of us was an ex-army officer who, it was said, had served in the 'Artists Rifles', a very odd military misnomer if ever there was one. He used to give us lectures on such things as how to pitch tents in a military manner, and how to dig the latrines, establish a cook house and, believe it or not, where to put the horse lines. We listened in amazement to what seemed to be so unlikely.

We also were instructed on the protocols for when we would be posted to some far corner of the Empire. Visiting cards or calling cards were an essential part of an Officer's equipment and we were told how these should be presented when we visited the local Resident Administrator or some other big-wig. Leaving one's card appeared to be a vital first step in being accepted in the community; it was very Victorian, old-fashioned thinking. From the way they taught, one did wonder if the Zulu War had yet ended.

One afternoon each week concluded with a cross-country run. Five of us mates, 'the ex', Dennis, Maggie, Ron and myself, a happy band of no-hopers, felt that this imposition put us to the back of the queue for showers and freedom to go to the cinema, or whatever we had on our program for the evening. It was obvious to us that the only solution was to get back to our billet ahead of everyone else. From the athletic viewpoint we were outclassed by most of our unit, therefore all that was left for us was to cheat. The next cross country we arrived home well in front by the simple expedient of short-circuiting the course. Clearly our instructors had doubts about the performance. The next week we were not so lucky as there was an N.C.O. at every checkpoint who demanded, 'name and last three' digits which he duly ticked off his list. Now we were even more determined to out-smart the system. So at each checkpoint we had one of us circle the checkpoint several times singing out the name and 'last three' numbers for all of us in turn. This worked very well and we were consistently in the showers first. Eventually a slight problem arose. A certain person amongst the flight had been an Olympic cross country runner and the fact that he was consistently out-classed by a bunch of lay-abouts caused the authorities to think he was not putting enough effort into his performance. He was a good sport, however, and held his tongue. It was quite a few weeks before we were found out, by which time we had frequently enjoyed being first in the shower and the pub so it was well worth it.

There was a club in Newquay called The Old Studio Club. This was an old fish loft and it contained a pinball table, a bar, a piano and a few tables and chairs. We were all members and spent many happy hours trying to make one drink last all night. Money was often an issue for us. Dennis preferred a café on such occasions where we could order a pot of tea and ask for more hot water for the rest of the evening.

One day Dennis had a date with a Miss Ainsworth from the Women's Auxiliary Air Force (known as the 'WAAF') and for some reason he was unable to make the dinner. He asked me, foolishly as it turned out, if I would go to the appointed meeting place and pass on his apologies. The lady concerned was very good-looking and apparently talented. Having conveyed Dennis' regrets I volunteered to escort her to wherever it was she had expected to go. From then on we went about together whenever we were both off duty. Marian Ainsworth was a concert pianist in civil life and a regular performer on the B.B.C. although for such events she used her stage name of Beatrice Ainsworth.

I was frequently out-ranked by various officers from all three services who wished to make Miss Ainsworth's acquaintance. As I said, she was beautiful and talented and for some strange reason she preferred me, an impoverished character at the bottom of the heap. Marian came from one of the upper class families with plenty of money. Despite this she insisted on living on her service pay, which was not much. Her grandfather was Richard Strauss, a world renowned composer, who was the son of Johann Strauss, and she grew up in one of the most musically talented and well known families in Europe. This made her part Austrian and explained her proficiency in German language as well as her brilliant musical talents. She got her name from another ancestor, Harrison Ainsworth, the author of such novels as *The Tower of London* and *Guy Fawkes*.

The day came when we completed the Initial Training and I was posted to a flying school, so I said farewell to Marian. We planned to meet again and wrote to each other regularly until I was posted overseas, and I believe she eventually settled in the USA. It wasn't a very serious relationship, and just like Janey and Lilly, no one was interested in making any serious plans or commitments at the very beginning of what promised to be a long war.

The Elementary Flying School, to which I was posted was the Miles Aircraft Factory at Woodley near Reading. I could hardly believe my luck. It was a civilian flying school and we were housed in the airfield hotel, which was a great improvement on the living conditions the RAF offered elsewhere. We each had a well-furnished room and excellent meals. We weren't allowed to forget we were in the service, however, and had physical training every morning but now we had no Pay Parades.

The Miles Company paid us every fortnight with money in a little envelope. It was five pounds a week, which was not a lot, but relative to most jobs, I suppose it was a pretty good wage. I had organised to have most of my wage automatically sent back to my widowed mother. So I never had much money at all. After the war ended, when I saw my mother, I couldn't believe what had happened. Mother, rather than using the money to provide for herself, had been saving it for me. She hadn't spent a penny and it was all sitting in the safe!

At the flying school, the students had been divided into two groups, one group would fly before lunch and have lectures in the afternoon and vice-versa. The aircraft we flew were Miles Magisters, low wing monoplanes, with better performance than the Tiger Moths that were usually used for training. The vast majority of WWII pilots trained on Tiger Moths because, being a tail dragger, they required so much skill and coordination to fly, they quickly made for an experienced pilot. However, for what they had in mind for my group, the Miles Magisters were considered the best introduction to what would follow—Spitfires and Hurricanes.

I was surprised how many of the aspiring pilots failed the course. A large proportion failed before they reached their first solo. These poor dejected people were sent on to become navigators or radio operators. Perhaps it should have been called the 'Eliminatory Flying School'. My instructor was a man by the name of Wenman. He was a very competent pilot and we had a lot of fun doing aerobatics, low flying among the trees and 'beating up' (very low swoops)—over his girlfriend's house. Another student came back and announced that they had done a very low pass that morning and the instructor had managed to break several tiles on the roof of her house. He was as mad as a hatter, but he was a good instructor. Wenman made me feel in control of the aircraft and as though I could master it, and he broke down any fear barriers through his own crazy behavior, and display of courage—or craziness. He was a natural teacher and instructed with very few words, mostly using a 'see one, do one' approach.

One morning we did a lot of circuits and then my Instructor said I was to go with the Chief Flying Instructor (C.F.I), Flight Lieutenant Trewby. This, I thought, is the time they tell me my services are not required; but such was not the case and I was sent off on my own— the first of our cohort to go solo! He just got out the aircraft and said, 'No no, sit there, and around you go again'. I didn't have time to get worried. I just taxied forward, running over the checklists in my mind, and making sure I was doing everything we had practiced the last time and suddenly I was airborne. I was doing it! In that moment, I understood where the expression 'the greatest high' must have come from. I was having too much fun to be terrified.

Miles Magister, a two-seater monoplane used for basic training with RAF Pilots, wingspan 33ft, maximum speed, 142mph.

Chapter Ten
Wings and Things

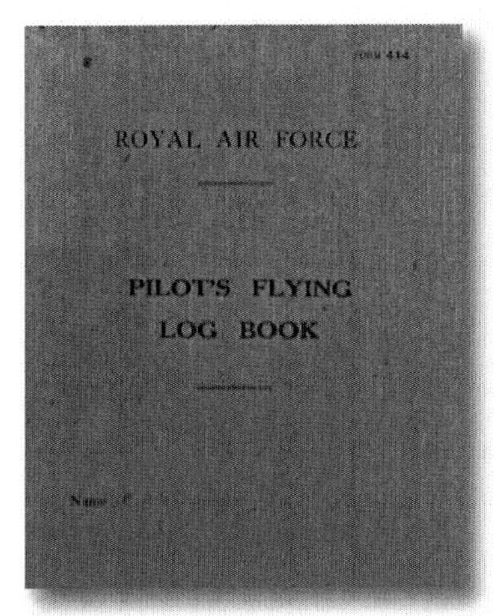

FROM WOODLEY I WAS POSTED to the Number 12, Service Flying Training Station at Spittlegate, Grantham. This was an old established Royal Air Force station and looked very permanent with its large two storey barracks, hangars, and various other specialised buildings, with which I would become familiar over the following months. The airfield itself was all grass, and the only paved areas were in front of the hangars. My friends, Maggie Carter, Ron Barnard and Ginger Jones, together with several others who I had trained with at Newquay and then at Woodley, were also posted here. This was marvelous from a social point of view, but I was not quite so pleased with the new accommodation arrangements. Being a service station our new quarters did not have the creature comforts of the civilian hotel, which was to be expected, but the food was definitely not up to standard. At Spittlegate things were much more regimented and this time we really knew we were in the Forces.

After the Miles Magisters we had been training on, the aircraft here seemed to be monsters; mostly twin engine Airspeed Oxfords and Avro Ansons, with a few Fairey Battles.

Fortunately they weren't anything like the Fairey Battle I had made when I was a boy, but even so, they had been taken off

Grantham 25 Course, May 1941, with Ron Barnard, and Maggie Carter front row, 5th and 6th position respectively, Ginger Jones and Cyril Johnson second row, 6th and 7th position respectively.

Front row left is Ron Barnard, middle is Maggie Carter. Second row, Ginger Jones is middle and Cyril Johnson is on far right.

the front line service, due to their poor performance in France, (in which I had already lost a distant relative). I was fortunate enough to be allocated to the group who would fly the Airspeed Oxfords, which I thought was a much more sensible aircraft.

After we were divided into various 'flights' we were introduced to our respective instructors. My instructor was Flying Officer Whitwell. The first day was wet and misty but nonetheless he took me and another cadet, Jerry Beasley, on a tour of Lincolnshire. I doubt if we ever got above 500 feet! My second lesson covered climbing, gliding and stalling, and my third instruction included single engine flying action in the event of fire and abandoning the aircraft!

Cockpit of an Airspeed Oxford.

Night flying lessons were more of a 'nightmare' by modern standards, because all flying was subject to the possibility of enemy intruder attacks. Consequently night flying was conducted from a bare field a few miles away. The rules were that if there was an attack, the flare path would be extinguished, though this in itself was no mean feat for the poor groundsmen, as it comprised of burning kerosene flares. Anyone in the air was to head east on

Period advertisement for the Airspeed Oxford, a twin-engined aircraft used for training RAF crews in navigation, radio-operations, bombing and gunnery. Maximum speed 192mph, wingspan 53ft, first flight 1937.

A Fairey Battle. Its first flight was in 1936, but by 1940 it had been withdrawn from combat service, and used only for training.

a prescribed track to a beacon, and then to orbit the beacon until the flare path was reignited. There were two problems attached to this procedure; firstly if you failed to find the beacon, as one of the students did, then you headed out over the North Sea. The pilot concerned happened to notice the efflorescence on the sea and headed back just in time to land safely. The second problem was that it was not unknown for the intruder to also orbit the beacon and shoot at whatever he saw.

Apart from the hazards for the poor student pilot, there were potential issues for those who were on the ground occupying the buildings that comprised of the night school. If there was any delay in extinguishing the flare path, the intruder might drop the odd bomb over the flying school. This happened one night and after the 'all clear' and the flying had resumed. An enthusiastic airman walked across the field and opened the door to the little flight hut, where the pilots were waiting, holding an unexploded anti-personnel bomb in his hand saying, 'Excuse me, Sir, but what shall I do with this?' No one answered but all left via the door, windows or whatever way they could. Fortunately I was not there that night.

One night, when I was safely tucked up in bed, I was awakened by a somewhat shattered man who burst through the door and told me that he had just been shot down. He was with his instructor in a Fairey Battle when they were attacked. The instructor rolled the aircraft and held it inverted telling the pupil to bail out. It was too late for the instructor who spun in with the plane and was killed. Experience had to be obtained within a very few hours.

We worked hard, and we played hard, with one weekend off every two weeks. On these weekends Ginger and I would hitch a ride to Nottingham, a city with a little more in it than Grantham. Getting back to base on a Sunday night was a bit of a problem as hitching a ride was unlikely. Consequently we had to remember to save enough money to get the last train if all else failed, which was not always forefront on our mind considering that Nottingham offered a lot of pubs, cinemas and girls. The latter had the reputation of being very obliging but Ginger and I never met any of those. We did, however, meet two girls when we went for a row on the River Trent. Two young ladies in a boat waved to us and we waved back. Then they kept waving and shouting so we rowed over to investigate. On closer inspection, we found that they had lost an oar and other than going round in circles, did not have a clue as to what to do, so we towed them back and went with them for a cup of tea in a nearby café. They took us to the cinema and we met them both on our next two weekend leaves. They were good fun, but I cannot remember their names.

An alternative, when we had a day off, was to visit my sister Monica who had a farm at a place called Knipton, a few miles away. Getting there and back was difficult and required a bus ride and a long walk. Mind you, in those days whatever we did seemed to involve a long walk, as the airfield was well out of the way of the normal bus routes and it was quite a distance into town. Still we considered it worth the walk to get some decent quality food as the slop they served at Spittlegate left much to be desired.

One day we were inspected more closely than usual and our uniforms had to be made presentable. We then learnt that the Station was about to be inspected by the Duke of Kent. We were instructed to carry on with our usual routines, in my case attending a signals class. I never did take to Morse code but I had learnt whom I should sit next to.

This man may have been good at Morse but by goodness, he was fearless in his approach to authority. The door opened and in came a parade of officers with much gold braid dripping off their sleeves. Most of them we had never seen before but we did notice our Commanding Officer. The Duke came around and had a diplomatic and encouraging word with each pupil and eventually he came to my neighbour. It was there that everything came undone for the rest of the gold braid brigade. The Duke of Kent asked what my neighbour had done before he joined the service. "I sold pin tables", came the prompt reply. It seemed that the Duke did not know what such a thing was. His education about coin slot entertainment began forthwith and I do not know how but the conversation quickly turned to food and the Duke's education was furthered by a very graphic description of the manner in which good quality raw materials could be converted into the inedible. Behind him the top brass were going frantic, but their glares and hand signals only drove my friend into greater detail.

After that I expected we would all be put in the camp dungeon, and if there was not one I expected it would be quickly constructed, but such was not the case. The next day there was a miraculous improvement in the catering. My friend was a hero although I doubt if his antics invited a promotion. Shortly after this the Duke was killed on his way to Iceland in a Sunderland Flying Boat, which was sad because he had impressed me as a very caring and interested person.

Part way through the course, a few of us were honoured by being moved to the large Officers Mess building, which housed the instructors and administrators. Here things were much more civilised. I was never sure of the criterion by which we were judged as suitable for the honour, certainly it was not on account of our being less like hooligans. I think there was an assumption of a better education. Of course there was a slight sting in the tail; we had some additional duties. There was also to be an improvement to our appearance and a tailor from one of the leading London retailers measured us for our new uniforms, which I thought promised of interesting things to come, but for the meantime it was back to our lessons.

One of the training devices was the Link Trainer, a dummy cockpit that accommodated a pilot and bomb aimer. In this simulator, a projector created a picture of the ground and the pilot's controls moved the projected image. The bomb aimer was then required to set up the variables such as wind, altitude, airspeed on the bomb sight and observe if he had hit the target. This was more difficult than it appeared. Another device to test the pupils was the 'camera obscura'. This was a darkened building with a lens suspended from the roof, which projected an image of anything that could be flying overhead. The pupil and the bomb aimer were required to track directly over the camera. The actual track of the aircraft was plotted on graph paper on the table. If you got the wind right it was no problem, but without that, everything would go amiss, and you would end up with a horrible graph as a momento. With all these sorts of exercises everyone had their turn at being the pilot, or navigator and bomb aimer, and we messed about and baited each other when we reviewed each performance.

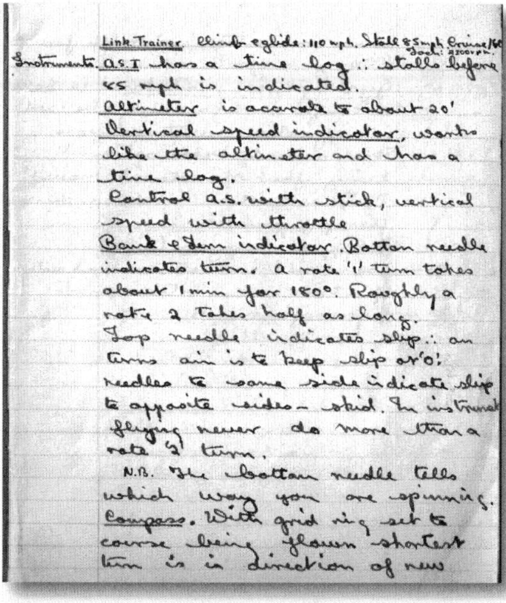
Link Trainer. Climb & glide: 110 mph. Stall 85 mph. Cruise 160
Instruments. A.S.I has a time log ∴ stalls before 85 mph is indicated.
Altimeter is accurate to about 20'
Vertical speed indicator, works like the altimeter and has a time log.
Control a.s. with stick, vertical speed with throttle
Bank & turn indicator. Bottom needle indicates turn. A rate '1' turn takes about 1 min for 180°. Roughly a rate 2 takes half as long.
Top needle indicates slip ∴ on turns aim is to keep slip at '0'. needles to same side indicate slip to opposite sides – skid. In instrument flying never do more than a rate '2' turn.
N.B. The bottom needle tells which way you are spinning.
Compass. With grid ring set to course being flown shortest turn is in direction of new

Cyril Johnson's Initial Flying Training course, first page of classroom notes, detailing the basic Link Trainer and controls.

In addition there was another exercise, the cross country, which really tested what we had learnt. The pilot was provided with a map on which a triangular route was plotted and no other information was provided. To fly that would have been the

A group of RAF student pilots during navigation classes.
Photograph published in 'The Royal Air Force in Pictures including Aircraft of the Fleet Arm', prepared by Major Oliver Stewart, 1941, p. 84.

proverbial 'piece of cake' if it were not for the fact that some pieces of the map were missing. If we calculated the wind correctly (this information was not provided) and maintained an accurate heading, there should be no problem, but people often stumbled their way through the exercise. A slight miscalculation and you wouldn't have much chance of finding where the next piece of the map joins up. I found that this exercise worked out rather well, on each occasion, which was fortunate, because I would certainly need these skills for where the RAF would post me.

Naturally there were a lot of skills to learn: single engine flying and landings, instrument flying, map reading, formation flying, precautionary landings, night flying, simulated forced landings, flapless landings, and evasive actions. Everything we were supposed to have learned was frequently tested and all this was crammed into a relatively short time, from the beginning of May to the end of July, 1941. When not flying there were still more classes to attend with subjects such as navigation, armaments, signals, and all sorts of other things which only the Air Force could have thought of. One of the strangest examinations we had was the night vision test which was very elaborate. In total darkness we sat and stared at a faintly lit screen and tried to make out what the images were. Equally difficult was writing down what we recognised with a pencil on a pad of paper, neither of which we could see. I suppose someone was able to read what we wrote for I was told that my night vision was above average.

Eventually, the course was over and we were all paraded and given a badge we would wear, with pride, above our left breast pocket. We had been assessed with regards to our ability, presumably to determine the most appropriate unit for the next phase of our career. I think I may have had the odd black mark against me with regard to some unscheduled flying (I wanted to find the best route to my sister's farm) which may have been the reason that my logbook says with regards to proficiency, "should soon reach above average", but at least they said I had above average night vision.

At the end of the passing out parade, the pilots were told their postings to the various Operational Training Units from which they would eventually be sent to squadrons. A few of us were left out of the list of postings and were told that we could go home on leave and would be informed subsequently as to where we were to report. My first reaction was 'What have I done now?' But then once I noticed that I was amongst those who had been the best performers, I felt things were not so bad, and in fact looking rather interesting.

I went home and was determined to have a good couple of weeks leave. My uncle Billy, who had replaced my retired grandfather, as the head village doctor, tried to conserve petrol, by attending his house calls on the horse and trap. He suggested that I might like to ride one of the horses, which he said was in need of exercise. The horse, Brandy, was duly saddled up. I should have known there was more to this than there seemed. Brandy was very big. I got on board and was about to turn his head in the direction of the drive when it became obvious that Brandy and I did not speak the same language. My knowledge of horses was very limited and I had no idea how 'the brakes' worked. We set off as if we were starting the Grand National Race. We headed for the tennis court, where we cleared the net and flew towards a little thicket, which led to the open field beyond. I hung on for dear life. Eventually when Brandy got tired we headed home

Cyril Johnson, in the back garden, the first time he wore his official uniform and 'wings'.

somewhat bruised by the low flying trees. That certainly caught me by surprise and I thought of how it would have sounded, given all the risks we had taken in our flying training, if I had been injured in my own back yard while on leave from the war.

There were of course the visits to relations but as food was so severely rationed, there were no feasts and as most of my friends were otherwise engaged on His Majesty's business, I was somewhat limited as to my activities. I did however make the time to see Lilly. We spent quite a bit of time together during those weeks, and each time I saw her I hoped my chances would improve. She told me that she would write to me every week while I was away at war, which I thought was very kind of her.

Proud mother Aida Monica Cooke, and Cyril Johnson in his new uniform, the day before he left for war.

Then she kept reassuring me about how I was a great friend, which was apparently girl code for something I may have missed. Eventually she told it to me straight; she explained that she would never marry me because we wouldn't get on well. I was gobsmacked! I hadn't even asked! We hadn't even courted! So that made for a rather awkward conversation; I wondered if I had been that transparent. But at least she did keep her promise, and true to her word, she wrote to me every week while I was away.

At the end of my leave I was advised that I was to report to the RAF Kemble, a place I had never heard of, somewhere in Gloucestershire. The day before I was due to report, I received my new uniform from Simpsons of Piccadilly, a successful English retailer who had only been in business a few years at that time. My uniform was a blue-grey colour, and because at this time I was still Non Commissioned, I had little VR badges (Volunteer Reserve) on my collar and now it was official, I had my 'wings', they were in gold thread and stitched above my left breast pocket. I was instructed to wear the new outfit so that I could have my photograph taken by my Aunt and my mother stood proudly by my side. She never said anything about how she must have felt seeing her youngest child in this guise, although I expect from what she had lived through already, she must have been terribly worried. Instead she enquired, again, if this was what I really wanted to do, and then said I should do it. It felt good to have her quiet sort of approval and I dared to believe that father would have been particularly proud.

Chapter Eleven
Young and Naïve

THE NEXT DAY, 9TH AUGUST 1941, I was taken to the railway station and installed in a first class compartment with instructions to change at Bristol. My luggage was stowed in the guards van by a friendly porter. When I arrived at Bristol and went to claim my luggage I was more than a little upset to discover that it was no longer on the train. Despite all the very clear labels as to its destination, it had vanished. There were discussions with the Station Master and the Rail Transport Officer, an army man, but there was no clue. The assumption was that it had been offloaded at some station en-route. The optimistic railway staff said it would turn up eventually but made no suggestions as to when or what I should do meanwhile.

There was little I could do other than continue my journey to Kemble and hope that it would turn up later. It did. Five or six weeks later, on the very day I was due to leave Kemble I was reunited with my trunk. In the meantime there was a problem because I only had what I stood up in; not even a toothbrush to my name. Like everything else in those days, clothing was rationed and to be able to buy anything one had to have the appropriate coupons, which I did not have. I tried my luck in the local town, Cirenchester, where I was at least able to persuade a local shopkeeper to let me have a clean shirt. A few days later the Adjutant, (a military officer who acted as an administrative assistant to a senior officer) supplied me with a couple of coupons and this got me a pair of shoes. Razor blades were not rationed so I was able to get a new razor and some soap. A frantic call to my mother brought me some civilian clothes, which I was able to wear in the evenings in the Mess, where I would not look too much out of place. Meanwhile I made whatever enquiries I could as to the whereabouts of this suitcase, as did the Adjutant, who did not know how to deal with a civilian in the Mess.

On my last day at Kemble I received a message to tell me there was a large suitcase at the local railway station with my name on it. I must admit, although I could see why everyone else thought it was a huge joke, my amusement was tinged with a little ill feeling towards the railway. The luggage fiasco was not the only problem that confronted me when I arrived at Kemble. The only money I had was what was in my pocket. Having received the Kings Commission, and having become a Commissioned Officer, my service pay was no longer paid in cash at a pay parade but was paid into an account at Lloyds Bank. That would have been fine were it not for the fact that firstly, the Air Ministry were in arrears in paying money into the bank and secondly, Lloyds had not sent me a cheque book. Once more, my dear mother came to the rescue and paid my mess bills and sent me some much needed pocket money. I did comment to our Commanding Officer that I thought it was a bit much to expect me to defend my country and at the same time, expect me to pay for the privilege of doing so.

Our Commanding Officer was an Australian who had been in the RAF for a fair length of time. Apart from a few peculiarities, he was the best C.O. I ever came across. Like my father, he never asked anyone to do what he would not have been prepared to do himself. Everyone on the station would have done anything for him and in fact they did. Every morning (except one memorable occasion) he would test the air with one of our aircraft. No-one else was allowed off the ground until he had completed his aerobatics and beaten up the station with his low flying swoops. He was a superb pilot. On one occasion he noticed that on the other side of our airfield there was a maintenance unit and in front of their hanger was the latest Mark Spitfire. Some airmen overheard him say that he would like to try it. When he came out to the field the next morning there in front of his office was the Spitfire, ready to go.

The C.O. was fond of a little drink before retiring to his bed. He also did not like to drink alone which, usually, was not a problem. One night, however, he returned from wherever he had been and found the bar shut and all his junior officers in bed. He went upstairs and up-ended the first bed he saw and deposited the occupant on the floor and demanded his attention in the bar. It happened to be me! That seemed like a good idea and so everyone else was awakened and the bar re-opened.

The Officers' Mess was an old and beautiful mansion some distance from the airfield to which we were ferried by a bus every day. The house was on a fairly large estate with a large ornamental garden and a great vegetable garden that supplied much of our food. Air Force men largely provided the labour for the garden. For the usual infringement of the rules, our C.O. would mete out punishment in the form of gardening duties. This seemed to be something that the 'Kings Rules and Regulations' did not prescribe but no one ever complained because it was not put down in their personal records. It also seemed to us the station Medical Officer (M.O.) never had much to do with people. It was evident that the M.O. was a keen animal lover, which led some people to say he was probably really a vet in civil life. This was reinforced by the fact that we had his parrot, dogs and horse on the premises. As no one was ever sick, nobody questioned his professional abilities. Kemble was a very odd place in many ways and the least like a Military Training Station one could imagine.

Another interesting character was an officer who, prior to the war, had given parachute displays in Allan Cobham's Air Circus. At the beginning of the war he had been instructing the first British paratroops and I suppose he came to Kemble for a rest, but what he was actually supposed to do there I never did find out.

Cobham's Circus used to travel round the country with his collection of aircraft giving displays and joy rides for a few shillings. The first flight I ever had was in his Avro 504 when I was 14 years old. That aircraft was considered to be a bit old, even in those days, and I am sure it would never have received a Certification of Airworthiness by the time the war broke out. It was also at Cobham's circus I saw a Pou du Ceil (Flying Flea). This Flying Flea had been constructed by some enthusiastic

Cobham's Flying Circus.

local, who had prevailed upon one of Cobhams' pilots to test it. Once airborne, the test pilot discovered that it would only turn in one direction and not in a smooth arc at that. Eventually he landed after a very wide circuit of the field. The machine had only one rudder pedal, the idea being that the rudder would normally give a turn in one direction if the foot was removed, because it was being operated in that direction by a rubber band. To go straight ahead the pedal was pressed part way and to turn in the other direction it would be pressed a little further. I think

A Pou du Ceil (Flying Flea).

the rubber band was a little weak. Furthermore, there were no ailerons as it was assumed the form of the wing would provide adequate stability. The theory worked, to some extent, but the response time was very slow.

But I digress. The parachute performer from Cobham's circus joined the Air Force to instruct the first British parachute troops and for a 'rest' he had been sent to Kemble. He was physically very well endowed and he did various stunts in the mess in the evenings from balancing a pint of beer on his head

Martin Herin, a performer from Sir Alan Cobham's Flying circus, wing walking, 1933. The only anchor between him and the plane is the rope he is holding in his hand.

and performing various antics without spilling a drop. Our C.O. who, as I have said was a trifle unusual, expected everyone to follow suit in whatever was going on. Our mess had footprints on the ceilings, just to show how aerobatic RAF members were without aircraft. Not content with such harmless fun there were other antics which would have been more suited to a circus. Then one night there came the performance to be remembered.

The idea was to dive over a chair, landing on one's hands and somersaulting onto one's feet. To do that over a normal dining room chair is difficult enough but over a high backed wicker chair seemed impossible. I think we all managed to get over the first chair but the landings were quite unlike the demonstration.

The end came when two wicker chairs became the hurdle! It was demonstrated by our parachutist. The challenge had to be met by our C.O. who, by this time was fortified by more than a drink or two. No amount of alcoholic stimulus would have persuaded the rest of us to attempt this circus trick nor any comments by our leader that we were a bunch of sissies. He went over the first chair with little trouble but the second chair proved to be his undoing. His head hit the chair and his landing was less than elegant. We failed to see him for the next two days as he was confined to his bed. We supposed he had been attended to by our vet/doctor and because professional ethics demanded confidentiality no one questioned the 'slight cold' official explanation for the pause in the morning aerobatic routine.

As clothing was rationed, and I still had no ration coupons, I would grace the mess wearing an old sports jacket and gray flannel trousers. This was a constant source of amusement to our eccentric C.O. and some of my fellow officers. One night I was set upon and de-bagged, after putting up a firm resistance. The result was that my trousers were removed in two pieces, two legs in fact. Now that would not have caused me much embarrassment in the course of ordinary events but on this night it so happened that one of the Officers had invited a lady friend. She had been busy sewing buttons on the said Officer's shirt, and came over to see what was happening. After the kafuffle I looked up, trouserless from the floor, and found myself looking at a startled female with a needle in hand. Hilarious to all the onlookers, not quite so funny for either the lady or myself.

Amongst the other aircraft, which had been collected at our airfield, was an old Avro Anson. The only purpose that it seemed to serve was to take important people for visits to other places. One Friday, our Commanding Officer reported that the weather did not look good for flying exercises that weekend and so he announced that a trip to London (via Croydon) would be a suitable alternative. No one counted how many people got on board the old four seater Anson, but at least a dozen of us managed to get there and back. We certainly never thought to question the meteorological forecasts for that weekend either. Years later I heard that this action caused the end of the C.O.'s

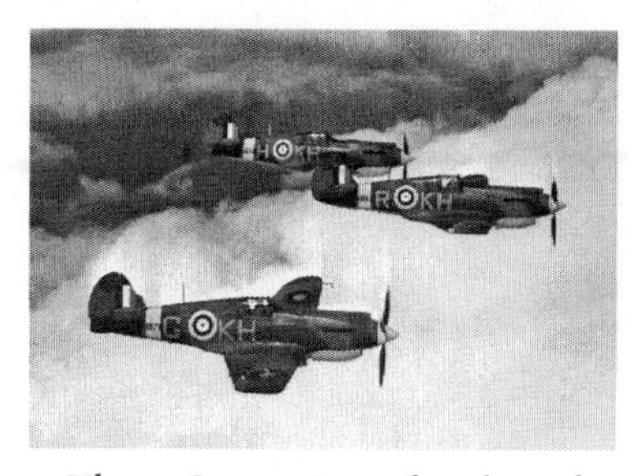

Three Curtiss Tomahawks (of No. 403 Squadron, based at Warwickshire [UK]), flying over the clouds in 'vic' formation. Forefront is a Mark I, travelling with two Mark IIs.

North American Harvard Trainer, maximum speed 208mph, 42ft wingspan (Image taken in Texas, 1943).

Hawker Hurricane—maximum speed of 300 mph and 40ft wingspan.

period of office. It so happened that on that very weekend the Air Officer Commanding decided on a surprise visit to Kemble and the highest ranking person he could find on the station was a Warrant Officer who apparently could think of no plausible explanation as to why he had been left in command of a virtually unmanned station, in fine flying weather.

We had a variety of aircraft to fly with a variety of peculiarities and complications. These included Harvards, Masters, Tomahawks, Oxfords, Hurricanes and a few other different Marks and models including the American twin-engine Marylands. Apart from the number of different aircraft we were to master flying, the complications came from the fact that some of the American aircraft had been made for the French Government and had a peculiar version of what we would call 'metric' instrumentation. Others were not even partially converted. Consequently airspeed indicators could be graduated in miles per hour or kilometers or even knots. Altimeters could be in feet or metres, boost pressures in centimetres of mercury or in pounds per square inch, and the instrument panels between the same type of aircraft looked different from one cockpit to another.

Airspeed Oxford, 34ft wingspan and maximum speed 192mph.

To make matters even more interesting, some aircraft had only some of their instruments changed from imperial to metric or nautical standards and sometimes a mixture of both!

One morning the C.O. asked me 'Did you learn French at school?' 'Yes, sir' I said, not having the faintest idea where the conversation was leading. 'In that case' replied the C.O., 'you should have no difficulty flying that Maryland'. He told me the speeds for take off, stall, how the fuel system worked and sent me on my way. He was not sure which instruments would be in what units. The Maryland was a slender machine with the navigator, when there was one, somewhere under the pilot's feet. When I was in the cockpit I found out why learning French

Martin Maryland, an American designed light bomber. Wingspan 61ft and maximum speed, 300mph.

Miles Master, a two-seater, monoplane advanced trainer with a top speed of 240mph, and 40ft wingspan.

at school was significant. However my school French did not include the technical language of aviation, not that I was ever very good at even basic French anyway.

With the aid of a mechanic, who seemed to be very amused by something, I got the machine started. No problems so far. Looking at the array of switches, which would have been more suited to a Wurlitzer organ, it appeared that few of them could be significant. I gathered that 'le train' could only mean the undercarriage and the other important things were obvious. From memory, I was told the climb speed was 140 k.p.h and landing 110 k.p.h. It climbed very easily and the wheels made the right sort of noises on their way up. While cruising round the circuit I noticed that a little red light was blinking at me. My immediate reaction was to wonder which engine was on fire but then I lowered the undercarriage and it went out so that solved that problem. Life was not meant to be easy and certainly not in the Air Force. About a week later I taxied along in the same machine and I was alarmed to find I could not get it up to 140 k.p.h. but the fence and trees were coming up fast! Then it dawned on me: they had changed the airspeed indicator to miles per hour!

After my nerves recovered from the Maryland, and the flight was successfully completed, I was asked to keep flying this wide variety of aircraft. I got the feeling I was being trained to be able to fly almost anything, with any instrumentation, at a moment's

Two Maryland cockpits, with a vast array of instruments to keep an eye on, and just enough differences to make flying a challenge.

notice. One minute I'd be flying circuits in a Hurricane and the next minute it would be a cross-country flight in a Tomahawk. The latter I particularly enjoyed because on one occasion when there were clear skies this gave me an opportunity to see what my home village looked like from the air. In the other direction I was able to do a few low passes over my Aunt's house in Droitwich, but in both cases no one stepped outside to notice me.

Towards the end of our time at Kemble, a small group of us, including Maggie Carter, Ron Bernard, a man named Jerry Beasley and one or two others, were told that we were going to Africa. Briefly someone explained what we would do there, but no one was ever told when or on which ship, or from which port we were to embark. I was excited to be going to Africa. Already from the snippets that they had shared, it sounded as though it would be full of excitement and real adventure, but more than that, I would get to experience what it was that father had talked about so passionately, and see some of the scenes I had dreamed of since I was a child.

As I had suspected, there was some reason for the eclectic collection of aircraft. We had been selected to be based in Africa on the understanding that we were bright enough to be able to fly a wide variety of aircraft—with instruments in English or French, metric or imperial and sometimes even both. We would just sit in these aircraft and work them out, either while we were on the ground, or in the air, but hopefully, before we touched the ground again.

After six weeks at the school, with very little experience, they were willing to write the following in my log book:

This is to certify that Pilot Officer Johnson has qualified to fly the following types of aircraft and is conversant with the fuel, oil and ignition system on each: Harvard, Master, Tomahawk, Hurricane, Oxford, Blenheim and Maryland.

I felt this was a gross overstatement and indeed, I tallied the following total flying hours of experience: Oxford, 30 minutes dual instruction, and 30 minutes solo, Blenheim 4 hours dual and 7 hours solo, Harvard, 45 minutes dual and 2 hours solo, Master 35minutes dual flying, and 2 hours 30 solo, Tomahawk 5 hours 30 solo—with no dual instruction. In the Hurricane I also

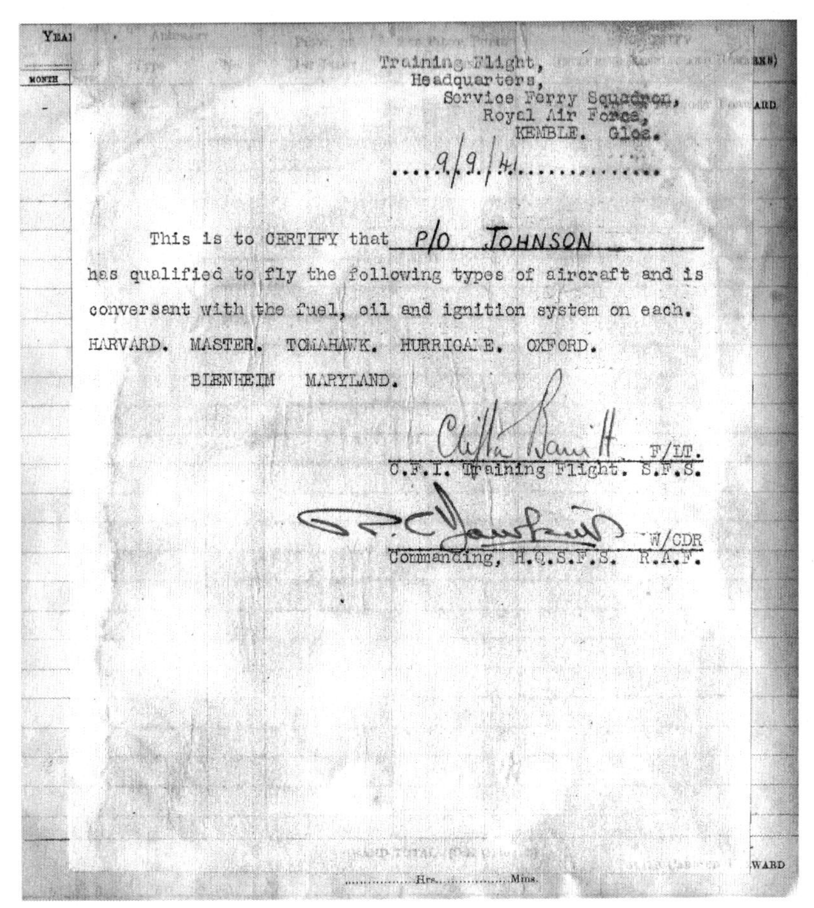

Training Flight,
Headquarters,
Service Ferry Squadron,
Royal Air Force,
KEMBLE. Glos.

....9/9/41...............

This is to CERTIFY that P/O JOHNSON
has qualified to fly the following types of aircraft and is conversant with the fuel, oil and ignition system on each.
HARVARD. MASTER. TOMAHAWK. HURRICANE. OXFORD.
BLENHEIM MARYLAND.

F/LT.
C.F.I. Training Flight. S.F.S.

W/CDR
Commanding, H.Q.S.F.S. R.A.F.

Excerpt from Cyril Johnson's log book, 9th September 1941.

had 5 hours 30 minutes solo. That was it, I was qualified on all of these machines. I really don't know how we got away with it! It all seems very naive to me.

Once more I enjoyed a few days leave at home, before the goodbyes started sounding ominous. Everybody told me to take care, and tried to warn me of whatever dangers they thought Africa held. Mother, on the other hand, was very good. She of all people knew not only of the realities of Africa but also of a World War and in particular she knew the stories father had told her of war in Africa, and yet she never said a negative word, and wished me well in what it was that I had chosen to do. My

AIRCRAFT	ENGINE
MAGISTER.	GY.P. MAJ. I
OXFORD	2. CHEETAH [illegible]
BLENHEIM I [illegible]	2. MERCURY XV
HARVARD	1. WASP.
TOMAHAWK	1. ALLISON [illegible]
MASTER.	1. KESTREL.
HURRICANE I	1. MERLIN.
MARYLAND.	2. TWIN. WASP
HURRICANE II	1. MERLIN [illegible]
BISLEY.	MERCURY.
WELLINGTON Ic	Pegasus XVIII
WELLINGTON III	2. XI. Hercules
HALIFAX V	Merlin XX
LANCASTER	

Log book record of the many and varied aircraft Cyril Johnson flew during WWII.

Aunt Teresa came and wished me well, and this time she and her son, Dennis, gave me a black leather bound diary. Teresa asked me to take good notes, and to promise that one day I would write a book on my adventures, a promise which I intended to keep, even if it would take me several decades before I started.

A few days later and I was in a transit camp at Hanforth, between Wilmslow and Styal Cheshire, not far from home and where my cousin Sylvia and her husband John Preston lived. This enabled my friend Maggie Carter, whom I had trained with from my first day at Newquay, and I to spend much of our time in the comfort of their home.

At that time there was a friend of Sylvia's staying with her who was the wife of a doctor. She was a good sport and had very amusing tales to tell; stories that made me realise that either all the world had gone mad, which seemed possible with the war, or that being innocent and naïve was no protection. We weren't kids anymore and the world we were growing up in was changing faster than us, which probably explained why my parents' generation were so often pointing out that our modern generation was not doing things the proper way, and starting sentences with that dreaded phrase, 'back in my day…' The anecdote I remember most concerned one of her husband's patients, a very young lady who complained of feeling nauseous in the mornings. When told by the doctor that she was approaching the state of motherhood she was not amused. 'How can that be?' she asked, for 'I am a virgin!' The doctor said he understood it had happened before, but that was two thousand years ago. She then admitted that there had been some heavy petting with her boyfriend to which the doctor said, 'The little buggers can swim you know'. A sign of the times perhaps? It certainly accounted for some of the public health messages and propaganda posters that we were seeing around England.

It was accepted, somehow, that many of the young boys who had left their mother's nest for this war, were still young enough to need lessons on the birds and the bees, and yet at the same time, mature enough to be sent overseas, fighting battles of life and death.

WWII propaganda posters aimed at reducing the rates of sexually transmitted infections.

Chapter Twelve

Following Father – the Voyage to Africa

IT WAS SAID THAT IF you frequented the local pub, the publican would tell you details of your voyage and in due course it so happened that he did. I don't think we were supposed to know where the train was taking us but thanks to the publican we knew our first stop was Glasgow, Scotland. It was a wet and miserable day when the train arrived at the River Clyde, but we weren't there long enough to grumble about it, as we were soon ferried out on a lighter, a small boat able to dock in the shallower waters, which would take us out to a large ship.

I set out on this voyage with starched new shirts, a hard covered blue log book which was still new enough to crackle down the spine each time I opened it and my leather diary, full of hundreds of fresh blank pages, promising of an adventure that would surely unfold. When I started writing, I didn't think my entries would come in handy for anything much, because although I intended to keep my promise to Aunt Teresa, at 21 years I was still invincible and believed that I would always remember the stories—in great detail and for all times. I wrote mostly because it passed the time and I only filled out my log book because it was compulsory and sometimes checked.

Seventy years later and I now know, that with time, even the most impressionable memories tend to fade, for better or worse. Holding a box full of diaries and flipping through their crisp yellow pages takes me back to these days, and somehow that which I did write reminds me of all the details, even the ones that weren't specifically recorded. One memory leads to another

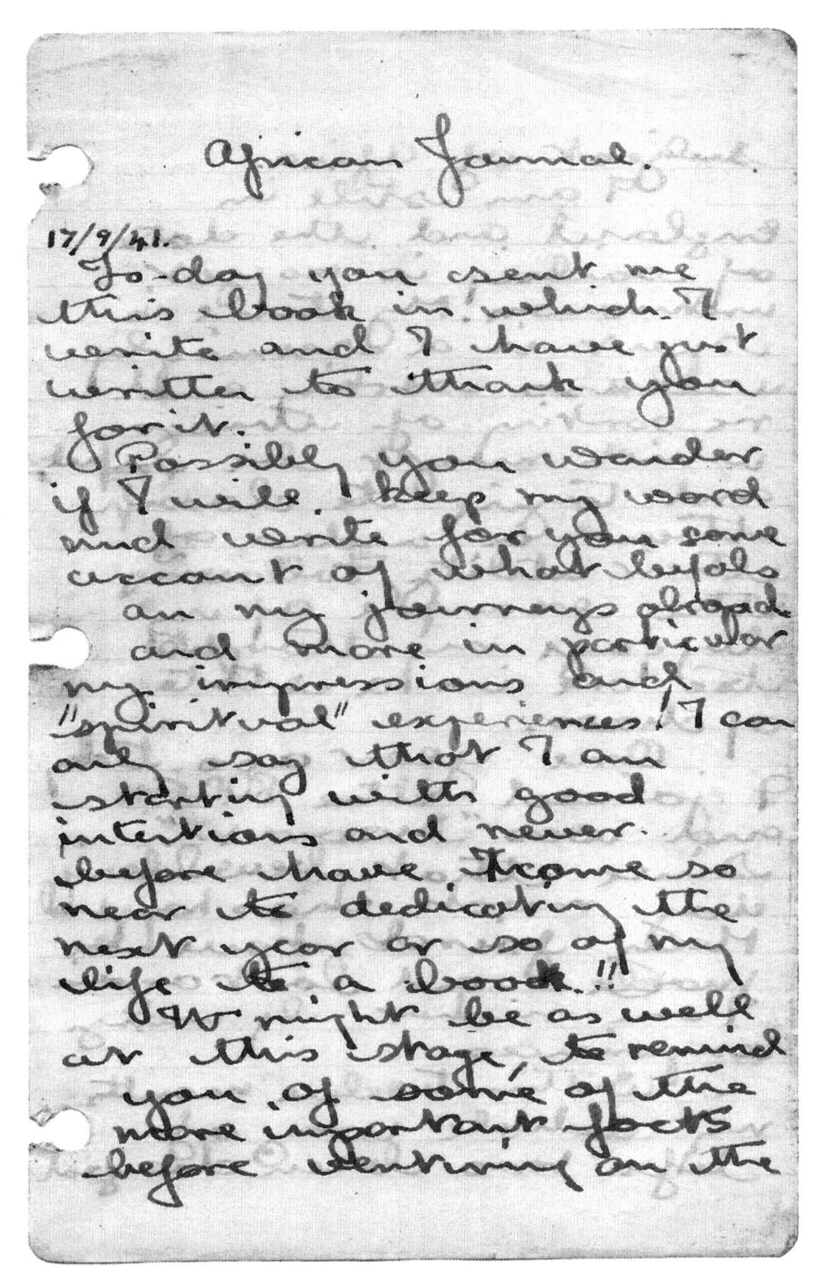

African Journal.

17/9/41.

To-day you sent me this book in which I write and I have just written to thank you for it.

Possibly you wonder if I will keep my word and write for you some account of what befals on my journeys abroad and more in particular my impressions and "spiritual" experiences! I can only say that I am starting with good intentions and never before have I come so near to dedicating the next year or so of my life to a book!!

It might be as well at this stage to remind you of some of the more important facts before venturing on the

The first page of Cyril Johnson's African diary.

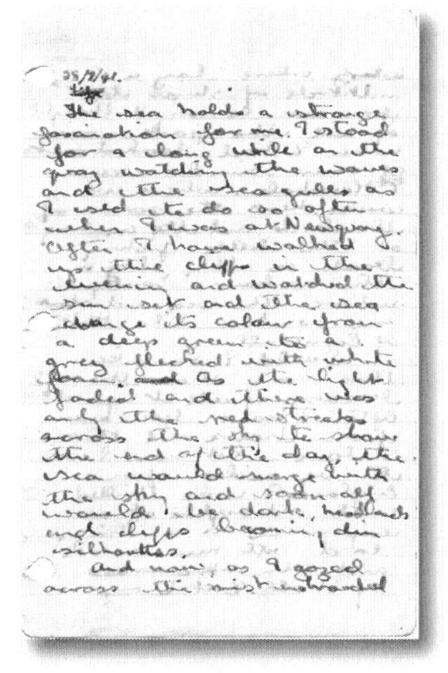
28/9/41.
The sea holds a strange fascination for me. I stood for a long while on the quay watching the waves and the sea gulls as I used to do so often when I was at Newquay. After I have walked up the cliffs in the evening and watched the sun set and the sea change its colour from a deep green to a grey flecked with white foam, and as the light faded and there was only the red streaks across the sky to show the end of the day, the sea would merge with the sky and soon all would be dark, headlands and cliffs becoming dim silhouettes.
And now, as I gazed across the mist-shrouded

Would I remember this more than the rest because of all it implied? After I have been away from home and those I love – but now I was to be exiled from England. I was thrilled with thoughts of new experiences in a strange land with new scents and the unknownness of it all. But in my heart was sadness, I was alone. I cannot recall all the dear memories that crowded into my mind, those places that I love

Excerpts from Cyril Johnson's Diary (reproduced in full below).

and the flow comes naturally, but I am certainly glad to have a record of all the names, dates and places and a few sentences about the activities of most days.

The following is what I wrote in my diary at the beginning of my voyage from the UK to Africa:

> 28/9/41
>
> *"The sea holds a strange fascination for me. I stood for a long while on the quay watching the waves and the seagulls as I often used to do at Newquay. After I walked up the cliffs in the evening and watched the sun set and the sea change colour from a deep green to grey flecked with white foam. As the light faded there were only red streaks across the sky to show the end of the day, the sea would merge with the sky and soon all would be dark: headlands and cliffs becoming dim silhouettes. And now as I gaze across the estuary where lay a multitude of ships which will form our convoy, I thought of all the past occasions when I had looked at similar scenes: Mousehole, where we could hear the*

scream of the birds and the creaking of the little fishing boats as they lay in the harbour, or Southport where the sea and shore are in no way like Cornwall.

Would I remember this more than all the rest because of all it implied? Often I have been living from home and those I love—but now I am to be exiled from England. I am thrilled by the thoughts of new experiences in a strange land with new scenes and the 'unknowness' of it all. But in my heart is sadness. I am alone. I cannot recall all the dear memories that crowd into my mind, those places that I love so well. I try to make memory of the trees in autumn, yes autumn is here now, and when I am about to leave the shore—the leaves are dry and brown, many of them are blowing along the lanes scurrying on their way. Soon the wind will be shaking the branches in its rage striping the few that cling to life, those that have the presumption to hold on when nature has decided that they shall fall. But I shall not see autumn through, I shall not walk alone, listening to their symphony and seeing them dance their ballet—for I am exiled. I dread, lest some unthinking person should come to talk to me, for I knew that the words would catch on the lump in my throat."

I was shaken from my thoughts by having to board the small lighter and take my luggage with me. This was a hard task as my, and indeed everyone's baggage, was far from light (as it included a 'camp kit' comprising a bed with a folding frame, a canvas wash basin and even a canvas bath!) I was carrying everything with me but the kitchen sink. When we were safely on board and had cast off, my thoughts began to settle with the rhythmic waves and I became aware of how hungry I was. I had not eaten since the day before.

Past ship after ship we sailed, every time saying this must surely be ours, as we headed out further we just kept on sailing past each of them. I wondered if the lighter was going to take us to Africa. Eventually we came to the *Narkunda*, the largest

The *Narkunda*, P&O Ship built in 1920.

P&O ship, built in 1920. The ship was enormous, over 600 feet long and nearly 70 feet high, about the height of a seven storey building. It was initially steam powered and converted to oil by 1927, boasting three towering funnels and some 15,300 horsepower, and from my brief experience on my uncle's horse, Brandy, that would have been a heck of a lot if one could manage to keep them all heading in the right direction!

Apart from its magnitude, my first impression was of menacing Lascar faces, militiamen from the Indian Subcontinent, peering over the rails at us. I was amazed at the incredible number of troops already aboard. I didn't like the appearance of the overcrowded ship at all! As it turned out, my impressions of forebodings were right. In November the following year, after disembarking almost 2000 troops, she turned about for home, and was bombed that evening. Fortunately only 31 people were killed, a number which would have been a lot higher if it had been hit only a few hours earlier.

After climbing aboard this enormous ship, we put our things in the cabin and had some lunch. This meal was very good and eagerly awaited. We were served by Lascars, for whom my original apathy was soon overcome by the speed of their service and the excellence of the meal. I positively beamed on them as I left the dining room. Rations didn't seem to be an issue and everything on the boat was as though we were still in peacetime.

Our cabin was on 'C' deck and about ten feet square. Originally it contained two beds. Now, however, they had another bunk over each so our cabin contained four; Maggie Carter and Ron Barnard who was also from Newquay, a South African named Adams and myself. Just down the passage, was the cabin of our other friend, Jerry Beasley, with whom we had trained at Kemble. I was glad that Jerry had been selected to go to Africa, he was a very precise sort of man, but still knew how to have good fun, and this at least meant that there would be a few familiar faces on this foreign continent.

I had plenty of time to write in my diary at this stage of the war and the following snippets capture the essence of the voyage:

29/9/41

"No move has been made and we still lie at anchor. The day was spent exploring the ship. Maggie and I were shown round the engine room by the third engineer."

30/9/41

"Rumour flies round the ship. This time it was right; we have sailed! Other rumours are about our first port of call."

5/10/41

"The routine of the ship is very simple. Meals occur at eight o'clock, one thirty and eight o'clock in the evening and as you can well imagine, I am always punctual. Soon I shall be very fat as I go right through the menu and often have the alternative dishes as well.

At ten o'clock every morning we have what is referred to as a 'muster parade' in which we all assemble at our various stations at which it is supposed, we should present ourselves in the event of our ship going down. One could hardly refer to muster stations as lifeboat stations for they have no lifeboat. True there are rafts, which are meant to be thrown over the side and we are supposed to swim to them. Of course there are some lifeboats on the ship but they will only hold 1500 and there are well over 2000 of us! So I hope the boat does not sink.

After this we retire to the grossly overcrowded lounge to have a drink and read before lunch. After that we sleep!

There are thirty nurses on our ship, who were divided out among the Navy upon their first appearance. One can only suppose the Navy is hard put to it. None of them look very young and most of them do not look very interesting. Three of them are quite passable and one, the youngest, is about thirty I would say. One of them seems to be doing very well with a Fleet Air Arm penguin. They are referred to as Romeo and Juliet by the unsympathetic onlookers. I am not in a position to criticise, but I don't like his uniform!

The Narkunda is very well equipped, especially with food we had not seen in Britain since rationing started at the beginning of the war. The ship sailed in a convoy, and we were able to see an aircraft carrier next to us and occasionally they hoisted up a 'Stringbag' [the nickname for a torpedo bomber biplane already outdated by 1939] on to the deck but we never saw it airborne. All things considered it was a pleasant trip along a route my father must have sailed many times."

9/10/41

"I have been walking round the ship after dinner tonight with Maggie and we noticed the phosphorescent glow in the sea, as if someone in the bow had sprinkled diamonds in the water. These little points of light show for a few seconds and are gone. They give off quite a bright light and some seem as big as saucers. So far

we have not been able to explain the phenomenon but it could be phosphorus left by fish or static caused by the ship. Maggie says he has seen it at his home town, Clacton, not in isolated spots but in long lines the length of the approaching wave.

Another remarkable sight can be seen during the day: flying fish! They are small silvery grey, about six or nine inches long with wings which look like those of a fly only much smaller in proportion to their size, about three inches long. Some of the fish have a double set of wings and are much bigger than the others. The distance across the sea which these creatures fly varies but I have seen one travel a distance which I would estimate to be about a hundred yards. I am told they obtain their flying power not from any flapping of their 'wings' but from the movement of their tails in the same way that salmon leap weirs."

Soon all the ships in our convoy slowed down, and one by one, passed through the booms into Freetown harbor, the capital of Sierra Leone. I was informed that the freed slaves, who had returned to Africa in the 1790's, thanks to the work of the British abolitionists, gave this town its name and I was told that the southern neighbouring country, Liberia, is also named as a part of the same liberation process. I wondered how it had been when my father called at these same ports thirty years before me and a hundred years after the abolitionists.

Had anything changed? Looking around me at all that was going on in the world, it didn't seem that humanity on the whole have come very far or if it had advanced by any measure, I wasn't sure in which direction.

15/10/41

"To-day we sighted land—just a faint outline in the distance that might easily have been mistaken for a cloud. We had been travelling in convoy in case

one of the ships was torpedoed by one of the U-boat submarines.

Naturally I was quite excited at my first view of Africa, Sierra Leone, and what struck me was the size of the river and the smallness of the town. Beyond the town the land rises up sharply into hills, which are wooded here and there. Scattered over the surface of the hills are several bungalows. The other bank of the river is flat and there appears to be a stretch of land behind which lie the trees without any signs of habitation. The town itself, I am told, is rather fine but small and with very little in the way of shops or amusements. To look at it I would say it was little bigger than our village.

Six of the party disembarked here, including the South African with whom we had shared a room, but I am told there will be no shore leave for us. "

18/10/41

"All day long for the last two or three days people have been coming and going to our ship in a variety of boats. Men and their baggage have been moved from our boat to the little Military Landing Craft or other lighters and rushed off. Officials have been brought in tugboats and launches, and most of them are being rowed by natives. About nine in the morning, other native canoes are rowed out and their owners dive for coins, which passengers on the large ships throw into the very clear water. It is a strange way to earn a living if ever I saw one, although watching them dive to such a depth is an impressive trick.

*The first man who came was a very expressive character and prone to swear as he paddled along the side of our boat saying 'shilling, sixpence' or 'anymore for me'. Of course people tried to fox him by throwing halfpennies wrapped in silver paper. At first he fell for it but when he discovered the trick he turned to his deceiver and said 'you f*** me up—you f*** off!' His* *price went up as business seemed to improve—'two shilling' and then to 'half-crown' before he would dive.*

I am told (and I can well believe) that the hardest part of this trade lies not in the diving but in re-entering the canoe afterwards. It is difficult because the canoe is so very light, being made from the bark of a tree and sewn up to a point at the bow which is finished at the stern by a block of wood. The sides are kept apart by wooded spacers, which are withdrawn to enable the owner to get in. Along the side the owner frequently paints in red the name of some ship he has seen. This colour seems to be very popular. The canoe is propelled by a sharp pointed paddle, which produces a fair speed, as the canoe is so light. Consequent to his frequent jumping in and out, it soon collects a lot of water and this is bailed out with a piece of wood shaped rather like the English domestic boiler shovel or alternatively he may push the water out by the dexterous use of his foot.

The 'bum boats' are numerous and they come alongside to tempt people to buy their oranges, bananas, coconuts, slippers and other leather goods. The price 'asked' is the same or more than that asked in England. I am told that ashore, you can obtain the same for about a quarter of that price. Having bargained and agreed a price the native throws up a weighted string which the passenger tries to catch. If successful you haul up his basket and put your money in it. This is lowered down and he puts into it your purchase. The 'bum boats' are usually of carved solid wood to the required shape and can carry a considerable load as well as several men."

20/10/41

"Today, Sunday, we left Freetown in what was formerly a vessel which plied between England and Ireland. It is much smaller than the Narkunda but with a similar density of population. On this boat

Portrait of Cyril Johnson in RAF Uniform

there are no fans but instead there are little blowers in the ceiling which have the desired effect if one can maneuver into their line of fire. Usually this is rather difficult as people come along and turn the nozzles onto themselves and one has to turn it round again when they are not looking. The food is rather poor and very limited; even a second piece of bread for tea is not allowed!"

I had finally made it to Africa, the place I had dreamt of in the stories father had told me. I remembered my youth in which I had dreamed of building bridges and bringing civilisation to Africa, and now seeing the vastness of the land and the incredible tapestry of culture, villages and markets along the coast, I began to realise that Africa had a life and rhythm of its own. Any schoolboy thoughts about just turning up and fixing Africa were by now completely dispelled, buried and never to return. We didn't have all the answers because we didn't even know what all the questions were, and it seemed rather ironic to me that we were the ones who were actively engaged in a World War. I had learnt from father that the influence of the west was not always a positive one, and that the most honest Africans lived inland, where they had never seen a white man; unlike those on the coast who had been taught how to 'trade' and with that, cheat. From what I had seen between our men and their encounters with the traders, and from the way people tried to drive such a hard bargain and play tricks on the coin divers, I thought this was likely true.

Chapter Thirteen
Darkest Africa

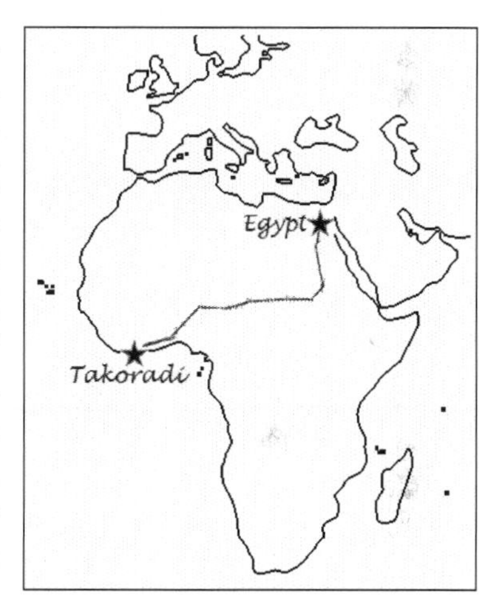

THE REASON WE WERE SENT to West Africa was to help the allies overcome the difficulties of supplying aircraft to the Middle East, particularly Egypt, where there was much fighting. Following the conclusion of World War I in 1914, Egypt had become a British protectorate. In 1940 the Italians had declared war on Britain and France and further complicating this was the fact that following the end of the First World War, Italy controlled large areas of North Africa, particularly along the North coast and around to Ethiopia. This meant that the enemy controlled much of the passage through the Mediterranean Sea. Furthermore, Germany now occupied France, so there was almost a complete stranglehold on the Mediterranean passageway. To have flown these aircraft direct from Britain to Egypt was too long a distance over either water or enemy territory.

The aircraft, mostly Blenheims and Hurricanes, were therefore shipped to Takoradi (in a country now known as Ghana) off the West Coast of Africa, and upon their arrival in crates they would be reassembled. From here, it was our job to fly them across to various bases in Egypt, which would save them from being shipped around via the Cape of Good Hope,

Takoradi Harbour, (now Ghana) as the ships ferried across aircraft and supplies during World War II.

South Africa, where many of the enemy's U-boats lurked. Apart from the dangers of sending the ships via the Cape, it also took seventy days for reinforcements to reach Egypt via this route. So it was clearly preferable for the allies to fly aircraft from West Africa to the Middle East, which typically took one tenth of the time.

Central sections of Allied aircraft are unloaded by cranes onto lorry beds where they will be transported to Takoradi Airfield and the full aircraft reassembled for its flight to Egypt.

The daring plan of flying aircraft across Africa came into effect a year or so earlier in July 1940, when an advanced party of technicians was posted to Takoradi. Their brief was to begin work on the facilities necessary to open the trans-Africa route, with the object of being able to handle up to 140 Aircraft per month (although its doubtful they ever reached this goal because one thing or another was always going

Hawker Hurricanes are uncrated and assembled for dispatch to the North African and Mediterranean theatres at Takoradi. In the foreground fuselages are pulled from crates, while in the background the completed aircraft wait by the runway to be ferried to Egypt.

wrong). The vast majority of the route was still unexplored, or at least unmapped, territory. By September 1940, allied air reconnaissance showed that the Italians had built up their aircraft in Libya in preparation to attack Egypt. There must have been a great pressure on the advanced party, for they worked so efficiently that the first reinforcement flight was able to take off on 20th September 1940. The strategic vision was a success. By October 1943, over 5000 aircraft had flown what became known as the 'Takoradi Route', something that greatly assisted the allies. However, this figure, though I am sure is accurate, makes it all sound a lot simpler than it actually was.

In 1941, a shaded political map of Africa looked like a dog's breakfast, with various patterns and blocks of countries settled or controlled by either Britain, France, Italy, Portugal, Spain, Holland or Germany, just to name a few. The route was like a hopscotch path if one was to keep flying above friendly territory. To make it across, the aircraft needed to be equipped with long range tanks, because there were limited places to refuel in the middle of Africa. It was indeed a very tough route from West Africa across to Egypt.

The route, which we were to fly time and time again, was accurately described in a book entitled *Royal Air Force 1939–1945 Volume One*; Chapter 10:

"The first stage, 378 miles of humid heat diversified by sudden squalls, followed the palm-fringed coast to Lagos, with a possible halt at Accra. Next came 525 miles over hill and jungle to an airfield of red dust outside Kano, after which 325 miles of scrub, broken by occasional groups of mud houses, would bring the aircraft to Maidugari. A stretch of hostile French territory some 650 miles wide, consisting largely of sand, marsh, scrub and rocks, would beguile the pilot's interest until he reached El Geneina, in the Anglo-Egyptian Sudan. Here, refreshed with the knowledge that he had covered nearly half of his journey, he would contemplate with more equanimity the 200 miles of mountain and burning sky which lay between him and Al Fashir. A brief refuelling halt, with giant cacti providing a pleasing variety in the vegetation, and in another 560 miles the wearied airman might brave the disapproving glances of immaculate figures in khaki and luxuriate for a few hours in the comforts of Khartoum. Thence, with a halt at Wadi Haifa… he had only to fly down the Nile a thousand miles to Abu Sueir. When he got there his airmanship would doubtless be all the better for the flight. Not so, however, his aircraft."

This route with all its stops, lines up exactly with my pilot's log book and diaries. My logbook shows that the trip took a total flying time of just over 20 hours. The reality was that it took at least four or five days, by the time we had made all the refuelling stops, and that was if all went to plan, which often it did not. I joined the North African scene, sixteen months after an advanced party, sent by the RAF, arrived in Takoradi and began organising such necessary facilities as roads, living quarters, mess halls, storehouses, medical support, gantries, workshops,

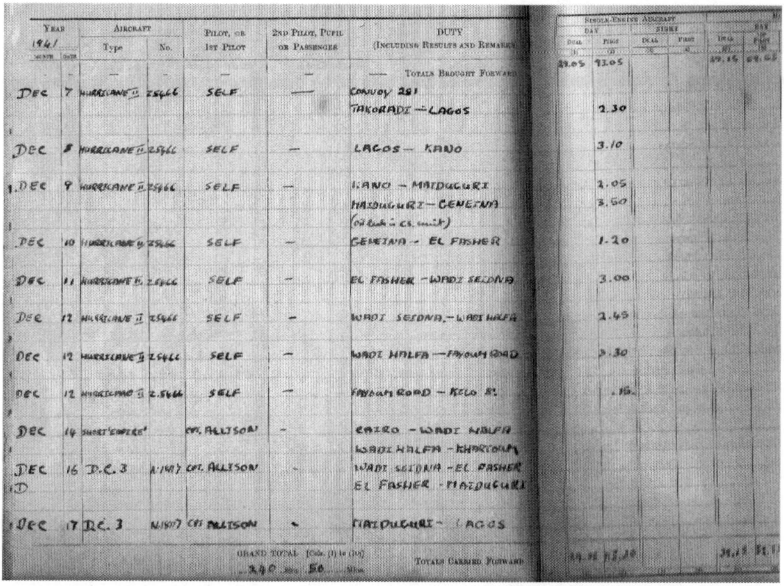

Year 1941 Month	Date	Aircraft Type	Aircraft No.	Pilot, or 1st Pilot	2nd Pilot, Pupil or Passenger	Duty (Including Results and Remarks)	Day Pilot
–	–	–	–	–	–	Totals Brought Forward	93.05
DEC	7	HURRICANE II	Z5466	SELF	—	CONVOY 281 TAKORADI – LAGOS	2.30
DEC	8	HURRICANE II	Z5466	SELF	–	LAGOS – KANO	3.10
DEC	9	HURRICANE II	Z5466	SELF	–	KANO – MAIDUGURI MAIDUGURI – GENEINA (oil leak in C.S. unit)	2.05 3.50
DEC	10	HURRICANE II	Z5466	SELF	–	GENEINA – EL FASHER	1.20
DEC	11	HURRICANE II	Z5466	SELF	–	EL FASHER – WADI SEIDNA	3.00
DEC	12	HURRICANE II	Z5466	SELF	–	WADI SEIDNA – WADI HALFA	2.45
DEC	12	HURRICANE II	Z5466	SELF	–	WADI HALFA – FAYOUM ROAD	3.30
DEC	12	HURRICANE II	Z5466	SELF	–	FAYOUM ROAD – KILO 8	.15
DEC	14	SHORT 'EMPIRE'		CPT. ALLISON	–	CAIRO – WADI HALFA WADI HALFA – KHARTOUM	
DEC	16	D.C. 3		CPT. ALLISON	–	WADI SEIDNA – EL FASHER EL FASHER – MAIDUGURI	
DEC	17	D.C. 3		CPT. ALLISON	–	MAIDUGURI – LAGOS	

Grand Total [Cols. (1) to (10)] 240 Hrs. 50 Mins. Totals Carried Forward

Example of a typical page from Cyril Johnson's log book during the time that he served in North Africa, showing the refuelling stops along the Takoradi route.

and hangers. They had done a lot of work in that time, given what little facilities they had started with, but there was still a lot more to do at some of the refueling stops across the central part of the continent.

> I wrote in my diary on the 24/10/41:
>
> *"After a most unpleasant trip we arrived at Takoradi, Ghana, West Africa. Maggie, myself and another officer went ashore with a view to picking out our own luggage from that arriving as the tenders brought it off. When we landed we were met on the quay by a most irritable Squadron Leader who informed us we had no business coming ashore as he had given express orders to the contrary (to whom he did not say, but it was not to us!) We managed, after much explaining, to tell him exactly what we were doing and persuade him to let us stay ashore.*

After a most exhausting morning doing nothing in particular, we were picked up and taken over to the mess in a van. 'Number Two Mess Takoradi' is situated on a hill and overlooks the sea. At the bottom of the hill lies a small golf course, which at first glance appears to possess a large number of bunkers. In fact these bunkers are the 'greens' made of sandy earth rolled hard. This is presumably because our English grass will not grow without weeds or will get burnt if it does grow. Outlined against the sea is a line of coconut palms, which is very arresting. They seem to stand out tall and slender with a palm bush on the top.

Our bungalows are arranged across the hill facing the sea with the main building in the centre. A further line of bungalows extends downwards towards the sea on the right hand side. My own room is second from the end of a block of ten. A verandah extends around all sides and all is raised up on a concrete base. My room is roughly square and contains a bed, wardrobe, table and an easy chair. To each leg of the bed is fastened by string a stick of bamboo, which supports a mosquito net. The former occupant of this room, recently gone to Cairo and is expected back any time, is a Pole. This is made plain by the presence of a pot of vanishing cream—only a Pole would leave such a sign of his vanity! The African 'boy' who looks after my room and me, is called by a name, which sounds like Alieza. He appears at seven in the morning with orange juice then disappears until seven in the evening when he presents himself for further orders which I usually fail to issue because I do not know what he should do.

The main building is made into two chief parts, the lounge and the dining room. The dining room has four long tables and overhead are three fans. There are two serving hatches hidden by screens fixed to the floor. In the corners of the lounge are long seats fixed to the walls, in the centre is a beautiful mahogany table and chairs and tables distributed over much of the remaining space.

Running the full length of the building is a verandah with wicker tables and chairs."

Number Two Mess Takoradi

26/10/41

"Rain has prevented our first day of scheduled flying in Africa. The rain came in torrents and was lashed by a fierce wind. Within a few minutes the drains were full and the road converted into a roaring river. Now the rain has ceased and the sun has gone down over the headland. The nightly chorus of insects has begun and keeps reminding one, even in one's dreams, that this is Africa. Home seems very remote now and all those I love are mere shadows in my mind. Yet they must all be there, perhaps someone is at this moment thinking of me. But I know so many little details of places and habits that it is easier for me to see them in my mind than it is for them to see me.

Here I have no fixed abode and am to move on soon. All is uncertain and unreal. Or is it that this is real and home a pleasant myth. It is very pleasant to recall all the days when England gave me what was best in her. Those days when we sat in the sand hills and drank a bottle of 'wine' with our food then lay in the warm caressing sun. Not the burning African master that sears the flesh and makes one sweat until one can sweat no more and you feel that you would turn to dust and crumple to a heap of ashes.

But even in England now the sun has gone behind the cloud and will only come out from time to time. Now there will be fog and frost, the air will be wet and all ones clothes penetrated by a searching drizzle. How pleasant."

My flying clothing and a few other things were presumed lost at sea. Maggie, Ron and Jerry thought it was hilarious that it should happen to me—twice, and suggested that I would likely be reunited with it on the day I would be due to leave, jesting that I'd have the newest looking uniform of anyone my rank. I thought it would present no problem as I could get a new issue from the R.A.F. clothing store at Takoradi: I was mistaken! The stores had nothing to offer but a pith helmet with a broad brim and a strap with which to tie it on and also it had provision for earmuffs. It looked like a leftover from the Boer War! However there were no earmuffs available and I was given a handful of cotton wool to put in my ears. For a short time, about five minutes, that might be effective, but of course I knew nothing of that at the time, and received it gratefully.

Perhaps by way of feeling sorry for me they also gave me the latest Air Ministry sunglasses. These were very heavy and comprised a frame attached to which were the sunglasses as a sort of visor which could be pulled down to block out the bright rays. When eventually I tried them and taxied out to the end of the runway, with the canopy open (as required by the RAF) all was well; but as soon as I opened the throttle and gathered speed the wind blew the visor down leaving me in darkness! That was the last use of those!!!"

27/10/41 I wrote:

"Down at the 'servicing flight' there was a Hurricane, Mk.I. The purpose of this machine can only be to play with. Today our Squadron Leader told us to fly it and do three landings each.

After several weeks at sea a pilot is apt to lose some sense of speed and his judgment of height is often impaired. However, I did not disgrace my self and my landings were quite good. Maggie was the first to perform and he scared the life out of us all by taking off almost in a dead stall, and then came roaring in emulating a scenic railway and finally ending his undulating approach by going round again. He surprised us because he can be depended upon as being a steady, safe pilot. However Ron Barnard and then Jerry Beasley proceeded to do much the same sort of thing so at least the Squadron Leader could not point a finger of scorn at only one (if that is any consolation) and any reproach was duly shared between all of us.

Fortunately the Hurricane was a fairly robust aircraft. I, wisely, as it turned out, did not attempt to get to the head of the queue and was content to watch a lot of other colleagues go before me. I noticed that there was a palm tree on top of the hill slightly to one side of the approach path. I considered that if I flew level with the top of the tree I would have a safe approach. It turned out perfectly easy. Just to prove it I did it a couple more times. Squadron Leader Holden was not impressed by the performance of my colleagues and sent most of them off as passengers to Cairo."

Holden then looked me over and his face revealed how unimpressed he was when he looked at my logbook which told him that during my course I had had only 1.3 hours dual and 15.5 hours solo on single engine machines; and 4.5 hours dual and

Takoradi bungalows, ten additional officers quarters near the main mess, where Cyril Johnson made his home away from home.

10.2 hours solo on twin-engine aircraft, making for a grand total of 31.5 hours flying experience and only 5.5 of these hours were in Hurricanes. I thought it best to hold my tongue, although the thought dawned on me that if he let me do a single ferrying operation up to Cairo, my experience in a Hurricane would increase four fold. Fortunately it wasn't necessary to offer such an optimistic opinion as Holden decided he would take a chance with me and after some local flying he let me loose on a convoy. The same went for both Ron and Maggie, whose performances of the day had been considered reasonable.

My other colleagues, including my mate Jerry Beasley, I regret to say, were dispatched onto passenger flights that would take them over to the Middle East where they could 'gain a little more experience'. How Holden explained that to his superiors I cannot guess. Fortunately, I did catch up with them later on during our service in Africa.

28/10/41

"Once again we performed in the Hurricane with a marked improvement with the exception of Ron who is apt to make good landings only after the first bounce."

30/10/41

"Ron has been told that he will be sent up [to Egypt] by passenger plane soon. In fact it was so soon he only had one hour to pack his bags and say farewell. I am very sorry that he should go and leave us, and as our sergeants are going too, it seems that there will be only Maggie and me left to hold the proverbial fort.

We have been given a Hurricane Mk II each and it is said we will join the convoy on Saturday morning. I have a feeling that something will come along to stop it—things do happen like that here in Africa."

1/11/41

"Today has been very hot and we have perspired from every pore! Fortunately we have been able to take the afternoon off but there was no transport going to Sekondi, the nearest town to Takoradi, and anyway the shops there would have been shut, so at least the afternoon was not expensive. But we were not able to buy any of the things we required. So Maggie and I went for a gentle walk round Takoradi and took some photographs. Later we went for a bathe in the sea. The breakers here are very large and one is often completely swept off one's feet. The surf riding is very good and it is quite common for the more expert to travel from far out at sea, right up onto the beach. I had mixed sensations as I looked up at the towering wave, which was going to crash down on me. For several seconds one was in a swirling mass of water, being thrown up and down and rolled over and over and finally emerged breathless and spluttering in white foam. The only objection to an otherwise perfect bathe is the large amount of seaweed

Aerial view of the runways at Takoradi, and barren surrounds in North Africa.

that seems to arrive and rushes round my legs with a most unpleasant tickling sensation and the sand gets in my eyes."

4/11/41

"Yesterday we were briefed for a 'convoy' to Egypt, of 1 Blenheim and 5 Hurricanes—the Blenheim's being the only ones with a navigator and radio operator. It did not depart, however, owing to bad weather. After a considerable delay we all retired to our various messes for the rest of the day.

Maggie and I took the opportunity of paying a visit to Sekondi, the local town. Sekondi consists of the Bank and the United Africa Company (Lever Brothers), which seemed to trade in everything. In addition there is an agglomeration of broken down native dwellings and a few small commercial enterprises. There was also a great little pub with a cinema out the back. It is curious that each store has a notice above the door stating that it is licensed to sell beer and spirits. Yet I did not see anyone drunk.

The stores sell almost everything and are especially well stocked with such desirable toilet requisites that cannot be obtained in England due to the rationing. Never before have I seen so many rogues gathered together in one place with the obvious intention of taking one's last halfpenny.

Africa has to be smelled to be believed. We decided to turn off the main thoroughfare and see some of the other sights of the town. At the end of a short street, Fifth Street I think it was, we came upon a large market place containing, amongst other things, a large building marked quite superfluously 'FEMALES'. The smell was horrid! The drainage system must be incredibly bad. It is odd that the natives who are so keen on having brightly coloured clothes should have ignored so completely the excellent perfume trade—and a great pity too!

Back in the main street we passed the Acropolis Hotel—with a home made cinema out the back. At the corner of the street stood a tall young native with his young sister (or was it his wife?). The girl was hardly more than fifteen and he was no doubt a few years older. As we passed they both grinned and the boy said "Massa, Massa you take my sister—very good!" But we hastened on, not really caring much for the sale and he, presumably, looked for a more likely customer. Later we learned from one of the men who had been stationed here for some time that this particular corner was their regular beat and that she only charged two shillings, but 70 percent of these women are diseased."

I enjoyed going to Sekondi, in part because I knew father had been there so many times in his ship. I would usually go with whoever was available at the time and keen for the walk and often this happened to be Padre Swallow, the military minister, who was a very realistic and humble man. I became good friends with him. His job was to help the men with a range of issues that came up, including issues from their families and letters from back home. He held religious services, which were very different to church back in England—mostly just about getting people together, and not all the 'religious' aspects that might have been expected. The Padre also had the task of censoring all the letters we wrote, making sure they didn't contain any sensitive information. He was, though I didn't realise it at the time, actually quite a hero. Though he never said anything of it, others informed me that the ship he was on when he came out to Africa had been torpedoed. Apparently while the ship was sinking he had kept many people calm and orderly while he directed them in lifeboats and when the ship sank he was amongst the last to leave.

Cyril's 1939 Kodak Camera

On one trip into Sekondi, with the Padre, we saw a young girl of about fifteen. She fell to the ground and had a seizure, and was frothing at the mouth. I was frightened for her. Naturally I wanted to do something to help, but the Padre informed me there was nothing that could be done. It was said to be caused by their beliefs in evil spirits. I was all for calling an ambulance, had we known who to contact, or if there even was such a service, which I doubted, but I wanted to do something at least. Still the Padre insisted that she was best left and instead he spoke to her family. It was very strange indeed, but it was an embedded part of the culture where people would put 'spells' on people. I would never have believed it if I hadn't seen it myself. Whether it was anything sinister or just their strong beliefs and the power of fear, such things often came true.

Chapter Fourteen

The Takoradi Route

LORD TEDDER, THE DIRECTOR GENERAL of Research and Development in the Ministry of Aircraft Production, reflected on how it was that several thousand aircraft found their way from the UK to Egypt, in his 1966 book, *With Prejudice: the war memoirs of Marshal of the Royal Air Force* stating;

"…. The route was by sea to Takoradi and thence by air on the Takoradi Route—that remarkable enterprise which was designed to play a large part in the build-up of Air Force strength in the Middle East…

… When widespread engine unserviceability was experienced, due to the severe operating conditions, this was to provide easy meat for hostile or ignorant critics. Navigation also presented a serious problem. The maps over long stretches of the route were practically useless, and the key was W.T. (wireless telegraphy) and D.F. (Direction Finding). One of the first convoys was lost (one Hurricane was lost and five scattered wide) due to failure on the part of an inexperienced wireless operator in the leading aircraft. The flow of aircraft was organised in formations of six fighter aircraft, led by a twin engined aircraft with a complete crew to serve as shepherd. We were very fortunate in the early stages to have the services of a number of very experienced Polish pilots, and they did grand work."

On one occasion Lord Tedder went as a passenger on a flight across the Takoradi route, and wrote in his memoirs:

"Before we took off from Takoradi, our pilot briefed me about the route and emphasised the necessary dependence on D.F. (direction finding) particularly on the stretch between Fort Lamy and El Fashir. He said that without D.F. that particular leg was

a gamble, and I was now able to see from the air how true that statement was. Nearly seven hundred miles of sheer nothingness; brown country, streaked with dry water-courses and dotted with bush; maps absolutely useless; nothing shown on them for the most part, for two hundred miles at a stretch, and where something was shown, it was obviously incorrect. I must say I would have hated to have to do that trip without a wireless."

Position lines by D.F. w/T.
The bearings obtained can be used as position lines. Bearings obtained in this manner are great circle bearings as radio waves travel along the great circle path.
Plotting position lines by DF w/T on maps
On all maps in use in RAF except those on mercator projection a great circle is represented, very nearly, as a straight line.
1) If Bearing of a/c from DF station is found the bearing may be laid off from station as a position line
2) If the bearing of station from a/c is found the angle of convergency ~~lines~~ of meridians between two positions must be applied to bearing before the reciprocal can be laid off

as a position line.
Plotting position lines by DF w/T on charts.
On charts (Mercator projection) a straight line is a rhumb line & a great circle is a curve on the polar side of the rhumb line.
The angle between rhumb line & great circle is the conversion angle & must be correctly calculated & applied to all D.F. bearings before being laid off on a chart from position of DF station

Cyril's notes from flying school on how to plot position lines on a map by Direction Finding.

That may have been Lord Tedder's experience, but I found that even the direction finding rarely worked. This should have been the time to put into practice the D.F. techniques that I had been taught in the flying school; that which seemed so simple when I sat in the classroom diligently writing notes. The theory was that one would use the radio to find a radio transmitter somewhere 'out there'. If they were able to find a transmitter, and if they could lock into its signal, they could use the instruments on the aircraft, to determine the direction and distance they were from the transmitter. If one was then able to lock onto a signal from a second radio transmitter, which was rare, then two lines could be drawn on a map, and where they intersected would be the location of the aircraft. You can imagine how useless this system was in a single seated Huricane, which didn't even have a radio! So this was yet another reason we travelled in convoy—with a Blenheim at the front as our guide. Not that they always had the right equipment either, but at least they did have a navigator and radio operator onboard who could usually figure something out—even if they didn't actually always have a radio. Indeed, radios were in short supply in those days and hence they were reserved for aircraft going into combat. You could only wave to your neighbor, and try to use various hand signals. Apart from this, there was no communication between the aircraft in the convoy; a real hazard, which was the undoing of many crew.

I agreed with Lord Tedder's sentiment when he wrote:

"My natural wish to study the working of the Takoradi Route was sharpened by the fact that the establishment of such a route had been and was a subject of considerable controversy. It was alleged, justifiably, that the establishment three years earlier of a weekly air service from Lagos to Khartoum was not a firm foundation on which to base a military reinforcement route involving for each aircraft some four thousand miles of formation flying. Much of it would be over practically unmapped country under conditions varying from tropical humidity to desert sand, and it called for repair and maintenance facilities along the route."

When we did finally manage to navigate all the way through to Egypt and safely deliver our aircraft, our journey was far from over. In order to repeat the process and ferry up more aircraft, we were shuttled back to Takoradi, in old out-of date machines, mostly Bristol Bombays. These were high wing predecessors to the Blenheim, but had had their hey day in the 1920s and so were considered pretty antique even at the beginning of the war. Many met treacherous fates. It was said by some, that you could navigate across Africa using wrecked Bombays as the landmarks. An exaggeration perhaps, but only just.

The British tried to purchase some 'DC3s', the American 1934 designed 'Douglas Commercial', a propeller driven, 30 seater airliner. It was thought that this would be the most efficient way to return the crews to Takoradi for their next ferrying convoys back up to Egypt, but the Americans refused to sell them to us. Instead, in true capitalist style, they offered the services of Pan American Airways—at a price, of course, and giving them a great dominance in civil aviation corridors across Africa in the years to come. There was a very interesting article that was published at the time from an American magazine that was pinned up in our mess hall as a joke—it stated how heroic the Americans were for pioneering a route across Africa, but in fact British Airways had been doing it years earlier!

The Pan American aircraft were flown across the South Atlantic from South America to Takoradi. To reduce the weight most of the seats had been removed and with that, the seatbelts. Passengers often had to sit on the floor, *unrestrained*. We were instructed to record passenger hours when Pan American carried us, which confirmed my suspicion that they were more concerned about profit than winning the war, a strong sentiment that also came through Lord Tedder's memoirs.

A Royal Air Force Hawker Hurricane Mk IIC (BD867)

On 3rd November 1941, I air tested a Hurricane MkII (BD783) and three days later I set off for my first ferry to Cairo, from Takoradi with Convoy 254 for Egypt. I never knew what 254 referred to but I doubt if there had been 253 previously so it must have had some other significance. The first leg to Lagos took 2 hours 45 minutes and it was then that I discovered the inadequacies of the cotton wool as a sound barrier. This was because I still had not been reacquainted with my lost luggage or issued a replacement helmet with the attached, protective earphones. I don't know if the cotton wool was supposed to be more of a gesture, but it was never really going to protect either my ears or my head. Being the last to take off from Takoradi, I waited on the ground, sweating in a very hot cockpit and when eventually I was airborne the cotton wool was a wet useless mess.

When I taxied to the parking area at Lagos I switched off the engine and an airman slid open my canopy and spoke to me. Despite his mouth being about 2 feet away I could not hear a word and told him I could not hear because of the engines that I thought were still running. He looked puzzled and looked round. Then I realised not a single propeller was to be seen in motion: I was totally deaf and the only thing I could hear was a Merlin engine in my head! After a day or two, my hearing seemed to get a bit better but it was never the same, for the rest of my life. Following that, I devised a better method of sound proofing my hat.

This trip was my first time flying from West Africa to Egypt, a trip I would come to do many times over in the period that followed. I flew the standard route from Takoradi stopping at; Lagos, Kano, Maiduguri, El Geneina, El Fashir, Khartoum, Wadi Halfa, and onto Fayoum Road—a strip of sand out near the pyramids in Egypt. The flying time on this first journey was the usual 20 hours 30 minutes that most flights would take. Much of the flying time was over inhospitable country varying from dense jungle to sparse desert.

The maps we were given were pretty useless. They were large fold up maps, but for all the details they did *not* portray, it would have made more sense to have just printed them on a postcard. Or in fact, they really could have said head east for two days and then northeast on the third and good luck finding the jerry cans out there! When the map came in its folded form, one could see reasonable details shown along the West coast of Africa. Then when the map was unfolded, it came as a great shock that for many hundreds of miles—in fact most of Africa—the next few pages simply had 'Unexplored Territory' written across them. By the time one got to the last pages, from Khartoum to Cairo, there was again a reasonable level of detail, showing a railway line along the River Nile, which we were to follow. Few other details were marked. During the time I was there they never reprinted the maps, or updated them, so I filled out my own details in the middle, adding to them with each journey.

The Hurricanes were each fitted with two extra 40 gallons tanks for the long hauls. An immersion pump enabled the petrol to be fed back to the main tanks. It did occur to me that if the pump failed life would be difficult! It was not possible to jettison the tanks and a 'wheels up' landing would present obvious risks. No one wanted to suggest anything other than to 'bail out', but,

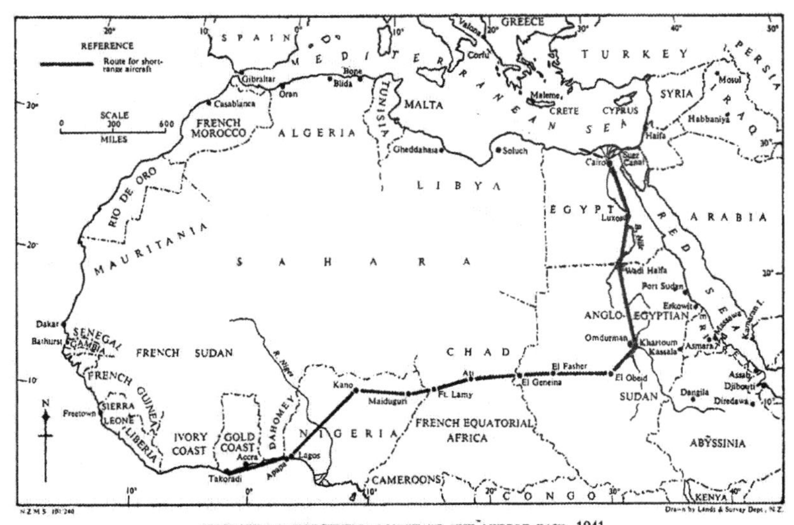

Takoradi Route—the map the RAF issued was in fact almost as sparse as this, with 'Unexplored Territory' written across much of inland Africa, and only slightly more details along the West Coast and the River Nile.

'down there', did not look particularly good country to bail out into, especially without a radio.

On this initial trip, I had my first encounter with the fauna of Africa. Upon landing at El Fashir I opened the canopy and there, with its two front feet on the wing, staring me in the face was a lion! I shut the canopy. Soon after a tall thin airman dragged the lion back and I was introduced to Leo who seemed to be a very large but well fed lion for which I was grateful. Lions are cats, and this one was no exception. He behaved just like a domestic cat. When Leo arched his back and rubbed against one's legs, you were well advised to brace yourself. A friendly lick was like being stroked with a rasp! Leo was incredibly docile and was particularly fond of a tall thin airman, whose method for moving him was to put the lion's tail over his shoulder and drag him away.

There was one incident, which I would never have believed had I not seen it. Petrol was delivered to the airfield by camels in cubic containers, the sides of which were about 18 inches. The

Leo the pet Lion, lying casually at the El Fasher airfield.
Note the unperturbed horse in the background.

cans were stored in a large pile just over 10 feet high. Leo was able to climb on top of the heap from where he had a good view of what was going on. On this day the airman, who I had met on my first encounter with Leo, walked past without seeing the lion; and I became worried when I saw the lion move to a pouncing position. Then Leo leaped off the pile and onto the airman. I need not have worried as Leo actually landed with his back feet on the ground and his front feet on the airman's shoulders gently pushing him to the ground. They both wrestled rolling on top of each other. I was told this was a fairly common occurrence, and the two of them had 'an understanding'.

As the war went on, more crews started to fly to Egypt via this route and it became clear that people were not happy having an uncaged lion roam about the place, even if it was a pet. It was decided that the lion would have to go, so arrangements were made for him to live in Khartoum zoo, in Sudan. To achieve this, the air force men at El Fashir made a cage to go on the back of a truck. Leo obligingly went in but then realised he had been tricked and created quite a fuss. His airman friend bravely went into the cage to calm him, and in the end, both Leo and the airman travelled together in the cage, a distance of about 450 miles, a full days drive, to the zoo. The airman was a rather unusual man about whom I wish I knew more. The only thing I heard after this was that whenever he had leave, he would travel to Sudan and visit his friend.

From El Fashir up to Egypt, the rest of my long flight was relatively uneventful. Our accommodation in Cairo was excellent. We lived on one of Thomas Cook's houseboats on The Nile, aptly named *The Egypt*. She was a large paddle steamer, which normally carried tourists up the river. Cairo was an exciting city to explore with shops selling all the things that had not been available in Britain since the rationing began.

In Egypt, I was able to visit Helen Massa, an old friend of the family, who now lived in Heliopolis, a tram ride from the city. She and her husband, Reg, were teachers at a school in Cairo. At this time Reg was away serving in the Army. Helen and Reg had lived in Egypt for many years and had a very comfortable flat in Heliopolis. Their household consisted of their five year old son, Paul, and an Egyptian maid Azziza and a man who looked after the place; a tall fellow from North Egypt. Paul was proficient in every language I suspect and was able to talk to the locals, in Egyptian, Italian, Greek or English. There were other inhabitants; a dog—a Boxer, which was fierce looking even with its filed teeth, and a couple of rabbits which I gather were there as part of Paul's sex education! The latter, I was told was frowned upon by some local English expatriates who believed sex was a very delicate, or should I say indelicate, matter not to be mentioned in conversation, and certainly not for children to have any idea about. They would have been received rather well in the village I came from.

While in Cairo I piloted a local ferry flight in a Harvard. Then I boarded a return flight by the Imperial Airways flying boat to Khartoum. It was very luxurious especially compared with Pan American, who took us the rest of the journey back to Takoradi where I arrived on November 22nd, after a round trip of 16 days. The whole process was then repeated with a slightly different end point in Egypt, leaving Takoradi December 7th and returning to the West Coast on December 17th.

December 7th 1941 was indeed a day to remember, for on this day Japan attacked Pearl Harbor. The news broke on the radio and everyone was talking about it! This altered things for Pan American Airways who had been flying the DC3s—because overnight they had become the American Army Air Corp and were soon reinforced with more pilots than aircraft. Some of them, having nothing to do, became temporary ferry pilots for the Hurricanes—six of them if I remember correctly. The first one did not do too well learning to fly the Hurricane and 'spun-in' to Takoradi harbour. The next one got to Lagos and landed 'wheels up'. I was there at the time and everyone could hear the warning horn as he gracefully landed on his belly. He said he could not figure out what the noise was. The aircraft was certainly in no position to take off again. I understand one got as far as Khartoum, but I am not sure how many, if any, made it to Egypt.

Meanwhile we continued ferrying aircraft up to Egypt and kept being transported back to West Africa in the American's DC3s mostly without seats or harness and our fares paid for by His Majesty's Government. Navigation was not a strong point of the DC3 crews. It was said that in the States they navigated by the radio beam, but across Africa this was not so easy. I was told that on one occasion a veteran RAF navigator being flown as a passenger in the back of the DC3, looked out of the window and failed to recognise the country side. Curiosity aroused, he went up to the cockpit to seek enlightenment. The pilot, who had always conducted his navigation by following a radio beam, remarked, "Gee dammit, I got the compass wrong way round!" Such errors for the pilot of a DC3 could be serious but nothing like the problems to be encountered when flying single engine aircraft with a significantly limited range.

CHAPTER FIFTEEN
Between Takoradi and Cairo

LAGOS, THE MAIN PORT AND most populous city of Nigeria, is a city about which I seemed to have found little sufficiently interesting, to note in my diary. It was hot, wet and smelly, although there were some hotels where one could slake one's thirst. I spent little time there as, generally after refueling, we moved straight on to Kano or Maiduguri. However I did make note of a story I heard from the RAF doctor who was based there, which I found to be incredible.

It seems there was a tough, adventurous expatriate lady who one day, while travelling up country in her car, ran over a native. When she got out of her car she found the victim was dead. It occurred to her that the next stage after being dead was burial. Accordingly, she dug a grave and interred the poor fellow. She then drove on to her destination. The next day she thought she ought to mention the incident to someone and possibly fill in some form or other. Accordingly she visited the District Commissioner and told him of the accident. The Commissioner asked 'Where is he?' When he was told, the Commissioner explained that she could not do that, and demanded to know more. Apparently, she then drove back and exhumed the body and delivered it to the Commissioner. More questions were asked. Apparently, the lady was potentially in more serious trouble to do with her illicit dealing in tin with a native at the mine, and her illegal investment strategies—the reason for her country drive.

Terrified by the possibility of imprisonment, in Nigeria no less, she wrote a letter of farewell and an explanation to her

A woman standing at the 500-year-old Kofar Mata, six-foot deep, indigo die pits.

husband. She then lit the fuse to a large amount of explosives under her bed. Her timing was very bad and her husband arrived home just in time to be blown up too. I was told the farewell note was retrieved in a decipherable condition, but neither of the couple survived. The exact points of truth about all this are difficult for me to say, given that I only heard the story third hand.

Kano, in Northern Nigeria was an interesting city. It was settled in the seventh century AD, making it the oldest, and now largest, native city in Africa. It was here that many of the caravan routes from distant places in the east, north and south converged. One of my many visits to the city was on 24th December 1941, and as our leader determined that the weather ahead was not favourable (and it was Christmas!) we remained there for the night. This was an opportunity to look inside the city, which is surrounded by a mud wall. I would guess that the wall was almost twenty feet high, twelve feet thick, and about a 15-mile circumference.

Two dyers displaying their finished indigo-dyed cloth with the traditional 'Bride and Groom' design.

Back in the eleventh century, when the wall was erected, the city gates were closed from sunset to sunrise, but nowadays, the entrance is always open and we were glad to find that it was just wide enough to allow a car to pass through.

One of the main industries of the city has always been,

A Photo Cyril took, inside the Kano city wall, December 1941.

dyeing cloth. The indigo dye is contained in wells about six feet deep and about three feet diameter. To dye the cloth its striking royal colour, indigo, potassium and ash are mixed with water and allowed to settle in the pits for a few weeks to ferment. The cloth is then submerged, and then pulled out and hung above the pit to drip off and absorb much of the pigments. It is subsequently submerged again and again to ensure that the eventual colour is as deep as is desired and the dye is permanent. Once it is declared to have been sufficiently dyed it is put over a tree trunk and beaten with wooden mallets until it becomes very smooth and glossy. The finished material, we were told, was often used in the manufacturing of hats. At other times, as they were dyeing cloth, they would tie it up in bunches with tight string to create traditional and ceremonial patterns.

To guide us around the city were two of the native policemen, who, if not respected, were certainly feared by the locals. If we stopped to take a photograph the policeman would ensure that everyone gathered and stood dead still, for the length of time that he determined necessary for a camera to capture a clear image; I took a photograph of a young girl with a tray of something on her head. At first I was surprised to see how many locals still wore white, rather than purple, but I guess with the temperatures there, wearing white made the most sense. It seemed that the purple clothing was mostly bound for export.

I noted in my diary, 24/12/41

"The alley ways are very narrow which means they are not well ventilated so one gets the benefit of the sweaty African odor. The stalls are constructed of earth pillars and if there is a roof it is usually of thatch. There is a variety of surprising items on sale: cheap mirrors, slippers (which the locals seldom wear), Christmas crackers, jewelry, 'quack medicines' and the like. People are buying, selling, fetching and carrying, and there are those who have become tired and fallen asleep on the road.

I am intrigued by the local version of a spinning wheel. An old woman sat on a mat and with her right hand she spun a top, attached to which was a ball of freshly shorn wool which she held in her left hand and thinned between the fingers of her right hand as it was wound onto the 9 inch long shaft, and made useful yarn, showing the great dexterity of the clever woman."

Photo of the local Kano markets, December 1941.

Christmas Day 1941

"It was determined that due to the harmattan—a dust storm which can be likened to a fog—the visibility at our next stop, Maiduguri, would make it inadvisable to proceed, and so we remained in Kano. Another Christmas with no presents and very few decorations to proclaim the day. Last night however, being Christmas Eve, there was a party at the local European Club to which we were invited. This did nothing to improve my vision however! While holding a drink in one hand and supporting the bar with the other I noticed a face that looked somewhat familiar. He was an Army Corporal and seemed to be looking at me and eventually he came over and asked had I been to school in Bolton. Then I remembered his name was Lomax and he had been at Thornleigh when I

attended. I hoped he didn't think this was the way I always behaved."

Three RAF men from Cyril's convoy trading at the local markets.

A few days later, we continued on our way to El Fashir, where Leo the friendly lion lived. But this time we were not greeted by a lion but by the C.O. His response to our arrival was to dash onto the dirt apron and say 'hurry up boys, get ready, we're going out tonight'. After that he promptly bundled us into his car and rushed us off to watch the local gymkhana motor sport racing. We were assured that all the people for miles around, who were of the slightest importance, would be there and looking at the well dressed spectators and the surrounding country, I could well believe it. Seeing the calibre of guests in our section was a wonder that commoners like us received an invite at all, and I was more than surprised to see such well dressed and important people attracted to such a contrasting, and irresponsible sport. It was essentially an obstacle course for cars, which had aptly borrowed its name from the equestrian equivalent and featured tyres, cones, barrels and the odd pile of rocks for cars to navigate, at high speeds while drifting on dirt roads. It was insane and I loved every minute of it!

Kano Mess Hall, during a harmattan (dust storm)
where Cyril spent Christmas 1941.

New Year's Eve was spent in Cairo where six of us celebrated by going to the Cairo Opera House which was a little like Covent Gardens and saw a 'Review' put on by the local British expatriates and the Red Cross. It was very good and highly entertaining. Afterwards we were invited to a party at the flat of a lady, on Gezaria Island in the River Nile, Central Cairo, where we had drinks, pulled crackers and did jigsaw puzzles with her 13 year old daughter before a typical English open fire. I may have missed Christmas but at least I had a New Year's Eve that reminded me of home, and suddenly I missed everyone back there so very much.

By January 2nd 1942, it was back on with the task at hand, and we flew British Overseas Airways Corporation (B.O.A.C.) to Khartoum, Sudan, where we spent the next day. Khartoum is situated at the confluence of the Blue Nile from west Ethiopia, and the White Nile, as it flows north from Lake Victoria, which makes it a little more hospitable, and interesting, however the mess hall here was still very primitive.

After a long flight, I went to use the toilet, a stand-alone unit, near the mess. I sat down, glad to be out of the sun for a moment, and breathed a sigh of relief from what had been a tiring day. There with my pants around my ankles, I was rudely awoken from my sense of privacy; there above the door was an 8-foot black mamba snake. In an instant, I recalled the legendary stories about Africa's deadliest and most feared snake, considering that they could: throw themselves from trees to strike at their prey, run faster than a man, and lift up to one third of their body off the ground in a near standing pose. The fear that came over me was enough to almost finish the job I had started!

It had seen me, and now stopped. I froze too, thinking of several things at once: the door opened inwards, my pants and belt were an obvious tripping hazard and sudden movements around a trapped animal were not good either. In a single movement, I managed to get my pants half way to my knees, held in one hand, open the door with the other and leap out, making a most spectacular and attention grabbing exit. Much to the onlookers surprise and then hysterical laughter, I danced around, half running, trying to my cover backside, and shouting 'snake' to those who were approaching.

I was still the talk of the mess that evening, but fortunately we departed the next morning for our next stop, where I hoped there would be less people to bring up the embarrassing subject.

I wrote in my diary, 4/1/42

"This night was spent in Maiduguri where three officers entertained us with a musical evening. The orchestra comprised a harmonium, a flute and an accordion. None of the trio appeared to be masters of their art as they would break off in the middle of a piece and condemn each others playing and then, like the mad hatters tea party, they changed instruments!"

Arthur Boddam-Wetham, a friend from the Initial Training Wings course.

After another overnight stop at Accra, now the capital of Ghana, I arrived back at Takoradi on 6th January and was informed there was a shortage of Blenheim pilots so I could expect a quick turn around.

My bearer brought me my orange juice and mail, and I opened a letter from a friend whom I had trained with on the first course at Newquay. My friend, Arthur Boddam-Wetham, whose family had all been Admirals and Air Commodores and the like, and who had always shown his kind and gentle teaching nature, had been made an instructor. In December 1941, while his pupil had been turning the aircraft around at the end of the runway, another plane failed to see them and careered into them. Arthur had been killed instantly. I was greatly saddened to read this letter and even more moved when I saw the little photo of him that was included. He was such a nice, young chap.

I put the photo into my scrapbook, next to a photo I had taken twelve months earlier, when three friends and I happened to pass through Lagos at the same time. In this picture stood two mates, Norman Steer and Siggy Oldfield, whom I met on

Four friends at Lagos Mess Hall, January 1941.
Jerry Beasley, killed in a Kittyhawk, March 41, Self,
Norman Steer and Siggy Oldfield, killed in Baltimore, Shabia, July 41.

Birthday card posted to Cyril, sketched by Lilly.

the Takoradi route, myself and Jerry Beasley who I had travelled out to Africa with and who was also from the same course as Arthur. Now seemed as good a time as any to make the updates to the caption, as Jerry Beasley and Siggy Oldfield had also been killed.

There was however some good news amongst my letters. Lilly had written to me and my birthday card had arrived a few days early. That cheered me up greatly, and it was nice to be remembered, particularly at such a time as this.

The next day I did an air test on a Blenheim and on 8th January, I joined a convoy of five Blenheims led by a Polish Squadron Leader. The weather was not what I regarded as suitable for flying anything, being very overcast with dark, low cloud. However our leader decided otherwise and we took off with myself as number five in the formation order. After taking off and circuiting the airfield number three appeared to break away and I never saw him again. We climbed on the approximate track to Kano, Nigeria, through the clouds, making slight turns now and then to avoid the thicker bits.

Something that every pilot knows is the danger of flying through cloud. Without the appropriate instruments, when you lose sight of the horizon, it is very easy to get totally disorientated and to veer off track. The only way to fly straight and level, is to rely solely on the instruments, and ignore your own sense of which way is which, in every dimension. At this stage we did actually have the minimum required instruments, they were a bit crude, but they worked well enough. If you had the confidence to trust them and let the readings override your own instinct, then theoretically you could have flown through cloud, although it was always avoided as much as possible. However, the main problem in this case was that we were flying in a tight convoy, and had no way of knowing where any of the other members of the convoy were without even a radio. Given how difficult it is to steer a straight course, in blinding cloud, the chances of a mid air collision were frighteningly high.

After about half an hour of weaving through thin cloud and fog and trying to avoid the thicker patches, the leader turned into a very large, thick cloud. If I held my course, it seemed to me I was in danger of hitting him, as visibility beyond my nose was non-existent in such cloud. So I made a slight turn to the left to keep away from him. At that stage we were over very hilly terrain, with lots of large rocky outcrops. As one might expect, when I cleared the cloud he was nowhere to be seen. I decided to maintain the track to Kano. However I soon met number four, Flying Officer Legge, coming back towards me! He too had lost the convoy. Under the circumstances, it seemed the wisest thing to turn around and follow him back to Lagos but I soon lost him too. It was obvious that I would have to find Lagos myself. I had no idea where I was, having not seen the ground since shortly after taking off and trying to navigate on my own under such conditions was incredibly difficult.

A short time later I considered I ought to be somewhere near the coast and I descended ever so cautiously, only to come out of the cloud somewhere near ground level and having no idea of the terrain! I hastily ascended to consider my situation. I can hardly tell of all that rushed through my mind in those few minutes, which felt like a lifetime. A thousand and one possible fates rushed to me and I was oppressed with a fear of something I could not define. I considered my options, and decided that if I could not see the sea soon I would return inland and bail out. At first this seemed logical until even more pressing thoughts took over my mind.

I could either roll the machine on its back and exit via the top hatch, or I could exit via the front floor hatch. To release that hatch I would have to leave the controls and climb forward, but the machine did not trim very well and when I tried to leave my seat she left the straight and level! If I could release the gun hatch I wondered how near the airscrews would be to my body when I jumped and what exactly was down there, below the clouds? Apart from that, the two propellers looked very close from this position! It was decision time. We had trained many times for bailouts, but nothing could have prepared me for how difficult it might have been to get from the seat to the exit. I fiddled with the latch on my seatbelt, and looked desperately out the window again, this time seeing a patch of cloud that was whiter and slightly wispier, but only just. Then there appeared a gap in the cloud and below was what looked like water. I knew I could descend safely if I was over water. This would give me the chance to see where I was without the risk of slamming into a mountain, tower or high pile of rocks that would bring my journey to an abrupt stop. I felt relief throughout my whole body, as I saw the water, and then I remembered to breathe. I wasn't safe yet, but at least now I could see where I was going and had the chance to orientate myself.

I was in fact over a small lake that I had never seen before. I followed the lake and then I beheld the sea. A further relief, except that a quick check on my map told me I was flying over Vichy territory, where the French were in collaboration with Germany. Flying just below the cloud, which was very low indeed, I flew over a Vichy airfield much to the amazement of the local gun crew who had no time to react. Fortunately!

I ascended into the cover of the white mist and got out of their view. Now, I had never been so thankful to see cloud and for all the invisibility it offered. From then on it was a reasonably easy matter to fly down the coast in just enough cloud to be well hidden, and at the same time be able to make out the coast line. Suddenly feeling exhausted and shattered, I flew along the coast to Lagos, where I landed without any of my convoy.

While waiting at Lagos to join the next convoy, and trying to recover, I was asked by a couple of Army Officers if I could take them for a local trip to view their antiaircraft and radio sites. Grateful for the distraction I obliged. These two gentlemen sat in front, one beside me, the other in the nose; what I did not know until after I had landed was that there was a third man in the turret. He had just climbed aboard and assumed I knew he was coming, not that it mattered much. I gave them several very low passes in different directions, which they seemed to enjoy judging by their expressions, and whether they saw what they needed to, or I was just there for their entertainment, I couldn't

Year 1942 Month	Date	Aircraft Type	No.	Pilot, or 1st Pilot	2nd Pilot, Pupil, or Passenger	Duty (Including Results and Remarks)
–	–	—	–	—	—	— Totals Brought Forward
JAN	2	‘EMPIRE’				CAIRO – WADI HALFA
						WADI HALFA – KHARTOUM
	4	D.C.3.				WADI SALDAH (Saidna) FASHER
						FASHER – MAIDUGURI
	5	D.C.3.				MAIDUGURI – KANO
						KANO – LAGOS
						LAGOS – ACCRA
	6	D.C.3				ACCRA – TAKORADI
						CONVOY 203.
JAN	8	BLENHEIM		F/LT. KIRCHNER	SELF	TAKORADI – LAGOS.
	8	BLENHEIM	29598	SELF	–	AIR TEST.
	9	BLENHEIM	29598	SELF	–	LAGOS – KANO (Leave convoy + return LAGOS)
	10	BLENHEIM	29598	SELF.	A.C. [illegible]	Local Reco.
						CONVOY 206
	12	BLENHEIM	29598	SELF	–	LAGOS – KANO
	13	BLENHEIM	29598	SELF	–	KANO – MAIDUGURI
						MAIDUGURI – EL GENEINA
	14	BLENHEIM	29598	SELF		EL GENEINA – EL FASHER
					1. PASS.	EL FASHER – WADI SEIDNA
	15	BLENHEIM	29598	SELF	3. PASS.	WADI SEIDNA – WADI HALFA
				GRAND TOTAL [Cols. (1) to (10)] 391 Hrs. 45 Mins.		Totals Carried Forward

Excerpt from Cyril's log book, January 1942.

be sure, but they certainly enjoyed the low passes and so did I. What I failed to see was that in one direction I had flown so low that I passed under the radio aerial, which they assumed was a good piece of RAF coordination.

On 12th January I attached myself to another convoy and resumed the journey to Egypt picking up a passenger at El Fashir on the 14th and three passengers at Wadi Seidna, in Khartoum, and then on 15th January I landed at Abu Qir, Wadi Halfa, famous for a naval battle in the days of Napoleon. I had suffered a toothache for the previous few days and it was not getting any better. One of the passengers was a dentist and a Wing Commander to boot! As it appeared that there would be delays in getting accommodation for the night, and further delays in being granted a warrant for the journey to Cairo, I decided to seek help for my aching tooth. The Wing Commander took a quick look and directed me to the Station Sick Quarters (S.S.Q.), leaving the other members of the convoy to sort out any paperwork and then pick me up. Darkness came before they arrived and I had not found anyone in the SSQ to treat me.

We tried to find accommodation in the Station but the Mess was full. We then checked the local hostelry, the Nelson Hotel. One would presume this was named after the victor of the battle of Abu Quir, Horatio Nelson, who routed the French. So it was a bit of a surprise to find that two French women, and a very disinterested spouse of one of them, ran the place. Perhaps the memories of that Gallic defeat, or perhaps it was, as they said and they did not have any accommodation, but whatever the reason, it was obvious we would have to look further and so we persuaded our truck driver to take us down the coast to Alexandria. The Squadron Leader and I sat in the cab with the driver and in the course of the journey we learned he was a German Jew.

I gave up nursing my jaw while we elicited his opinion of Hitler, the Nazis and the war in general. From his remarks he seemed to me to be a defeatist until we realised that through his imperfect command of English he frequently confused his negatives and affirmatives. Asking him if he thought the Allies would win the war he replied 'no' which filled us with pessimism until he added, 'Germany cannot last much longer'. To every question he gave the wrong answer until he qualified it. He would have made an excellent diplomat or modern day politician!

The best hotel in Alexandria was said to be the Cecil, but as it too was full we had no opportunity to verify this. The next place was the Metropole. When we entered we were met with a large notice reading, 'Officers Only'. The N.C.O.s in the party quickly removed their tunics except for one who was Irish and who had to be restrained before his sense of injustice could give the game away. The accommodation was sumptuous with a very large bathroom with a bidet. That was a puzzle for the radio operator, the fellow I shared the room with, and he considered what it could possibly be for. He thought it must be for washing his feet but that proved a failure with more water over the floor than over his foot, and any way, as he said, he could only get one foot in at a time.

Diary Entry 16/1/42

"My birthday—at least I remembered and in consequence took myself out for a drink and a seat in the cinema. Twenty two years old! When this war started I was nineteen and studying for part 1 of my Final Engineering Degree—I am now twenty two and have still a year to go for that degree and have in fact several wasted years between part 1 and part 2. When will I be studying part 2? These are supposed to be the best years in one's life but to me it looks like a void—a chasm into which I have fallen and wasted part of a lifetime!"

Diary Entry 20/1/42

"Four of our pilots had been missing for a week and it was feared they were killed. However they arrived back safe and well despite an unfortunate experience. All four were in a Boston, piloted by Freddie Rotheram with Taylor in the nose and Norman Steer and 'Williams' (a South African) in the rear. I had met them all from time to time in the various mess halls dotted across Africa and shared several drinks with them, but beyond that I didn't know any of them particularly well. They were flying over the rugged territory between the area of Port Sudan and the Nile. The pilot noticed the petrol gauges indicated the tanks were almost empty although he thought they had taken off with full tanks. He was unable to decide if there was a leak or if the gauges were faulty, or both. He decided that while he had power to manoeuvre he should land. They said it would have been a creditable

landing had it not been for a boulder, which took out the nose wheel. Three of them rushed out to the front of the aircraft expecting to see Taylor, who was ejected from the nose of the aircraft, likely dead or discover the extent of his injuries, but to their surprise he was quite unhurt.

For a day or two they remained by their machine to await their rescue. They tried to send a call out on the aircraft's radio, which was intact. Unfortunately they had not been given any information as to how it worked and they could not receive any incoming messages as they had no headphones. They decided the only hope was to turn the radio on and send a distress message in the hope that it would be received. As it turned out their call had been picked up, but the people at Port Sudan declared that the Boston did not have a radio so they did not search the position given!

The first sign of life was an old man with a small herd of goats who would have nothing to do with them, refusing to sell either his goats or milk. They considered taking the goats regardless but that would have been a desperate move that was fortunately discounted by the arrival of more friendly locals. They were given goat's meat, and water which was wet but very muddy. Offering promises of great rewards, they were taken to Port Sudan by camel, but little progress was made on the first day due, they said, to unfamiliarity with their mounts. After getting used to their mounts they made better progress."

They were lucky, and at least I didn't have to add anything to Norman Steer's name under the Lagos photograph, but for every story like that, sadly there were many others whose endings were not so fortunate. I wondered as to the possible fates of the four that I had taken off with only two weeks ago. People did go missing quite frequently. It could easily happen that there would be an aircraft problem, which meant that the pilot would do a forced landing. Most aircraft did not carry a working radio, so if the aircraft had been either lost or separated from the convoy and needed to make a forced landing, no one would know their location. Sadly this was not a rare occurrence. Many landed safely and were totally uninjured, but due to the damage to their aircraft they were unable to take off again. Such crews or single pilots soon found they had nowhere to walk to. There was no way to call for help. They could be in the middle of the African desert, often hundreds of miles from anything on their maps.

In one case the crew were found some time later. They had landed uninjured, but with no water or food, they had all shot themselves.

Chapter Sixteen
The Land of the Pharoahs

1943 US Air Force photograph of a Dakota air transport plane flying past the Pyramids of Giza, Egypt.

EGYPT WAS NOT AT WAR with the Axis, although the forces of the Allies and at times those of the enemy, Germany and Italy, paid little attention to the Egyptian border and the hot ancient land was an interesting country in interesting times. Her population seemed to comprise of every nationality but mostly civilian Italians, who ran many of the businesses, Greeks,

A local fisherman delivering his catch to the wharf markets.

Armenians, French, British and Sudanese, with the latter seeming to have a monopoly on the domestic security services for some reason. Yet strangely, although we were in fact at war with the Italians, here in Egypt, there seemed to be peace between the local British and Italian civilians. I was surprised how well I was received in many of the restaurants and shops even in my RAF uniform.

After a long flight across Africa, we would spend a few days in Cairo before being sent back to Takoradi to repeat the process. Here in Cairo we always enjoyed our down time in the city. It was a thriving place, and despite the war, there were a great many tourists and plenty of entertainment options including cinemas, which showed all the latest flicks from Hollywood and an Opera house, which made good use of local talent in addition to the occasional performers from both sides of Europe. How the stage battles weren't more than theatrics I don't know, but in the guise of their stage costumes, they must have forgotten that in any other context they would be out there fighting each other.

The Egyptian version of a Horse and Cart bound for market.

In the cities at least, the Egyptians appeared to have taken the Western enterprise system to heart and do their best to turn everything into a good profit. We soon discovered that, like West Africa, petty crooks, willing to do almost anything for an extra coin, populated the ports and wharf markets, but at least here the shops were well stocked, there were no rations and the food was plentiful and good. The fish markets were incredible, with some of the largest fish I have seen for sale, and the food was transported by means of a camel and cart.

The first night of one particular trip to Cairo I met a man, whose name I cannot remember. We decided—as we had only

200 piasters between us (then, the equivalent of 2 British pounds) —we would visit the local cabaret. As neither of us knew anything about such things, it seemed like a good idea. Upon arrival, we were met by two dark eyed maidens, who introduced themselves to us as soon as we had found a seat. The first lady asked me for a cigarette and I, unthinkingly, obliged her. Before I had a chance to consider things, she was comfortably installed in a chair next to mine and was doing what I suppose was her job as a hostess, which seemed to consist of calling the night-shirted gentleman to get drinks for us, and then for herself and the second lady! The vendors would then shell peanuts for us. Despite the playful way of the girl who had designs on me, well my money I expect, I did learn a few words of Arabic and she a few of English. Her method was to hold up some object such as a glass at which time I would say, 'glass' and she told me the Arabic word. It was obvious from their barrack room expressions that these girls had learned their English from the British military and so when we told the peanut vendor to go away, one of them added, 'yes, f*** off you bloody bastard' in an unfitting, neutral tone. Ordinarily that would not have surprised us, but coming from well-dressed members of the fair sex, in such an innocent context, we were surprised and very amused. When we told them we had no money left they quickly departed. I never discovered if the establishment paid these girls or if they lived off the commission from the drinks they had us buy, but with each drink they were given a small metal disc, which I suppose they would later be able to cash. Towards the end of the night I noticed that while we paid for wine, or whatever it was that they asked for, they, in fact, were only served with soft drinks.

During a layover on one trip to Egypt, there was a chance to visit the pyramid of Cheops and the Sphinx, which was only a short distance away by tram. My old friend Maggie happened to be up there at the same time so we took the opportunity to explore the ancient empire. The trams were always crowded and it wasn't unusual to have to let one or two pass before finding one with enough room to stand on. Our tram did not go the full way and we had to change and with this we were assisted by some helpful Egyptians, who tried to help us hail the exact tram, and even tried stopping a couple of taxis. However, their assistance was not without some expectation, including offers for us to buy their oranges or have our shoes shined. Undoubtedly, our uniforms gave the impression that we were able to tip well. So with oranges in hand, and four very shiny shoes, we found the right tram that ran down the central strip of a long straight modern road.

We had been in Egypt a few times before although we were not surprised when we were approached by an 'official' guide—albeit that his status was self proclaimed. He was very polite and had a perfect command of the English language. 'Before proceeding', he said, 'I would like you to see this little shop with its collection of genuine antiques found in and around the great pyramid of Cheops'. We were shown a large collection of alabaster pots and figures and scarabs, which varied in size from a quarter of an inch to about two inches long. Each had some meaning and often a hieroglyphic inscription, many of which were translated for us. One in particular, I was keen to buy, it was Shu the god of the air. It was carved from stone, no more than half an inch high and he was holding up his arms as if he were holding up the atmosphere. I do not know if it was his appearance or merely our commonality of interests that made me make friends with Shu but with every minute that passed as I examined him he became more attractive. They were asking ten shillings for him which I would have paid had I had the money. Unable to buy, we went on to look at the pyramid.

The guide told us that the pyramids were exactly square with faces north, south, east and west. (I failed to ask the obvious question, was it the magnetic or geographic points of the compass? Seeing as the magnetic compass rotates a few miles each year, I'd expect it was the former, in which case the puzzle of how they did this, becomes even greater.) The pyramid he said was solid except for the small chamber in the centre and the access passage. The pyramid was once covered with alabaster but that was removed for use in the Blue Mosque in Cairo. The purpose of the pyramid, we were told, was purely as a sepulchre or tomb for the ancient kings. Although I rather think that it was so that their names wouldn't be forgotten.

Near the pyramid are two smaller pyramids made for the Pharaoh's Prime Minister or the Pharaoh's family and the remains of the temple wherein lies the 'bath' in which the bodies were embalmed. On the east side there has been a cradle cut in the rock which housed the Pharaoh's ship. The stones for the construction were mostly imported, some of the granite came from Wadi Halfa which was an incredible feat in itself, but other stone came from a few miles across the river. Our guide told us that Prince Edward, the then Prince of Wales, climbed to the top of the pyramid with a golf club and drove a ball off the top. As the guide said, it would made a good tee. I am not sure if this is true or not, as it certainly isn't recorded in any books I ever came across, but it does make a good story for tourists.

The Flying Boat *Challenger* on her moorings.

After spending a few days in Egypt, the return trip to Takoradi was again by the British Overseas Airways Company (B.O.A.C.) Empire Flying Boat as far as Khartoum and thence by D.C. 3 by Pan American. This time, we had stayed overnight in a comfortable hotel and in the morning we put our luggage in the foyer as we had been instructed, where it waited to be collected and taken to the airport. I noticed a porter with baggage, including mine, which he was loading on a truck. I assumed Pan Am was being very helpful. I was wrong! My luggage, including my diary, was put on the wrong aircraft and went to South Africa. Fortunately, it caught up with me a few weeks later, and

Maggie, who had never lost his luggage, didn't fail to point out the hilarity of the matter, not that I agreed with him, of course.

Luggage for me was always an issue, but mercifully when I did lose it, it was never much at a time. Because of the nature of the job, we only carried the clothes we could pack into the limited space in the Hurricane—about three changes of clothes. The Blenheims were much better in respect to luggage stowage capacity. Although despite the lack of space in the Hurricanes, I soon found that I was able to pack quite a lot of loose individual items—in places such as the spaces around the eight machine guns. Given that I didn't have or need a lot of clothing and personal items, I usually tried to carry fruit such as bananas and pineapples for the staff on the remote staging posts whose diet, though adequate, was often a little boring. It was definitely not advisable to carry cans or bottles of beer or other such liquids as they could easily burst at altitude, but a few pineapples around the base of each machine gun seemed to fit comfortably.

Altitude was not often an issue, except for one memorable occasion when Maggie and I ended up in the same Convoy. He was flying in a Hurricane along side of me. We didn't use bottled oxygen when flying along the Takoradi route, perhaps it was considered unnecessary to fly so high, or more likely and as was the case with radios, this too was in short supply and so saved for those in combat situations. On this particular flight, we went high to get over the top of thick cloud, and suddenly I saw Maggie peel off to the left. He was heading downwards in gradual circles. I knew something was wrong, so I followed him down, not sure what I was going to do when I got much lower. He didn't seem to have any control over the aircraft. Then, fortunately, at only one or two thousand feet above the ground, he recovered and flew straight and level, and then quickly climbed back up. Indeed he had been unconscious, and said he 'just woke up' to find himself there, and thought he'd better do something.

Talking to Maggie later that evening made me realise how dangerous much of what we now considered 'normal' really was. As far as wars went, flying planes in Africa, was pretty safe; it was not as bad as the risks some of the blokes in the Army had to face, and certainly not as bad as being stuck in some ghastly trench, but there was still a lot that could go wrong at 30,000 feet. I didn't usually think about it very much; most flights went to plan and were very straight forward, and when they didn't there would be little time to ponder the dangers either, and usually I would be too frightened to be scared and too scared to be frightened. A pilot just had to pull himself together and do whatever was needed in order to get out of the situation in one piece, but watching Maggie, drifting down, felt as though the inevitable was playing out in slow motion before me. It was terrifying to see, and he was a very good pilot. I alluded to this briefly, but he did not engage in the subject and just went about his evening routine. He had a quiet confidence about him, which I admired and he was the only man I ever met who literally got on his knees to pray each night before getting into bed. As I tried to sleep, I thought it may have done him good for today, but would it be enough to see him or any of us through the war?

Another hazard was the lack of trees or significant landscape features to serve as references for landings. Misjudging height on a hot and glary sand airfield with no references was a common problem! Ron Barnard, who was with me from the earliest training days made a bumpy landing at Wadi Halfa. He was flying a large Maryland but the bump was serious enough for one engine to completely fall off! I saw him a couple of years later at Command Headquarters in London, when he had recovered sufficiently to fly a desk!

A further dangerous aspect of flying across Africa, that was often underestimated until it was too late, was the effect of large changes in temperature throughout the day. At certain times of the year, the West African Gold Coast can be very humid and one feels as though one is never dry, although the temperature is not as extreme as it feels. Interestingly, in the north of Nigeria or Sudan, a two day flight away, the summer heat is very dry with very low humidity, and although the mercury reads higher than in the tropical regions, one can get about in these conditions with less discomfort. However, whatever the type of noonday heat, the early mornings were often very cold. This change in temperature could be very disconcerting and dangerous.

One would set off just after first light, when it was cold, and they were suitably dressed for such conditions. Then for most of the flight at higher altitudes it was still cool, but once one came down for landing, it would get much hotter. This was usually ok, but the combination of a pilot wearing warm clothing in the hot afternoon sun, together with fatigue or any illness could be disastrous and people literally drifted off to sleep.

Apart from the heat, it was not unknown for accidents to happen for no obvious mechanical reason. One of our pilots arrived at a Staging Post and while on the circuit suddenly dived nose down full power into the ground. He died. Likewise, my friend Ginger Jones, who also trained with me from the earliest stages, flew into a tree for no apparent reason and he remembered nothing about the cause. Yes, he did survive, but it must have been a miracle because the fuselage of his Hurricane had twisted almost 360 degrees.

In the case of both pilots, investigations were carried out. Despite the efforts of the medical staff to maintain our health, many of us had life threatening issues. A customary blood test was insisted upon one day prior to commencing each flight ferrying operation up from Takoradi. It would usually determine if anyone had malaria, but there was no way an attack or relapse could be predicted. These were particularly tricky for the pilot to identify in themselves as they frequently started with being fatigued and shivering, followed by feeling very hot; all of which was consistent with early morning starts in the desert and noonday temperatures in a glary cockpit. Although they usually felt worse than a terrible case of the flu, a malaria attack could sneak up on a person and if the initial symptoms were not noticed or ignored, it could quickly manifest in neurological symptoms, varying from fatigue and decreased concentration to seizures and loss of consciousness. None of which is good for an airborne pilot! It was believed that malaria was the reason for Ginger's aircraft diving into the tree, and the other pilot's mishaps.

On a lighter note, attacks of the proverbial 'gyppy tummy' were common and there might also be no warning of the onset. The condition was unpleasant but sometimes amusing for other people. One pilot flying between Khartoum and Cairo was overcome by an unstoppable urge. What I saw was a Hurricane doing some very erratic formation flying. The poor fellow was trying to keep his cockpit clean. It seemed to him a good idea to undo his harness and remove his shorts and place his map and charts on the seat so as to receive the byproducts of his problem. Success achieved, he carefully folded the maps and opening the canopy, mid air, disposed of them. That was where he went wrong; the slipstream removed the papers but, unfortunately, the contents were blown back into the cockpit.

Chapter Seventeen

Sickness and Mosquitoes

APART FROM THE CHALLENGES OF being at war, and of living in a foreign continent, we had many health problems and diseases in Africa, which we knew little of back in the UK.

On 7th February, 1942, I wrote:

"Sunday. I have been feeling tired which is not unusual for Takoradi. My friend, the Padre, said I should see the M.O. [Medical Officer], Scott, who forthwith took me to the sick quarters where he measured my temperature and did a blood test. He issued me with instructions not to fly tomorrow."

Followed by another entry:

"Monday 8/2/42. I feel much better, very well in fact. The M.O. now thinks I can fly tomorrow but he will let me know the results of the two blood tests."

Poster used by U.S. Army Medics to remind people of the threat that mosquitoes pose, to them personally and to the outcome of the war.

The medical officer had explained that what he was testing for was a 'slippery' disease that tends to 'hide' so it often goes undetected with just one blood test. So now I had to wait to hear back on my second blood test, to confirm a negative result, before they would let me fly again.

By 9th February I wrote:

"Tuesday. Down to the flight office this morning feeling very well; never felt better! Norm Steer had arranged for me to join his convoy. Then I was told to go to the medical room for the routine temperature and blood test. During the proceedings the results of last nights tests came in and against my name, written in red, was the word 'positive'. I would have believed it on Sunday, but today? The Padre, despite my objections, drove me down to the European Hospital with a couple of books to read, 'if I felt up to it,' he said."

Apparently the test for malaria had come back positive. I informed the Padre that I was more than up to reading a book, and in fact I felt quite fine to fly, but he said, 'Well you most likely won't be up to reading tomorrow then'. He must have known a thing or two about the course of this particular disease; certainly much more than I did. The next day was nasty, to say the least.

Malaria is an infectious disease transmitted by the Anopheles mosquito, and the only way to protect against it is to avoid the possibility of being bitten. Not that I knew it at the time, but near the beginning of World War II, it was documented that across West Africa, the men in the RAF suffered from this at a rate of 843 per 1000 per year! In fact, it was so prevalent that they needed to increase the small local hospital to 300 beds to cope with so many people getting the disease. Fortunately, with meticulous attention to fly screens, bed nets, sprays and other repellants, the incidence rate was reduced to 78 per 1000 per year, just four years later and only 30 of the hospital beds were set aside for malaria. As a part of the RAF learning and attempts to reduce the incidence of this dreaded disease, during 1942 a lot of research was going into anti-malarial measures, including testing the best methods of dispersing insecticides from the air with low-flying aircraft. What this was supposed to do to our lungs, one can only imagine, but presumably it was thought better than having malaria. In the beginning of 1942, although they hadn't quite perfected the preventative aspect, there was at least a well-established treatment regime in place; so when I was finally ready to be released from hospital, and could sit up for long enough to write, I noted the following about my initial experiences:

> *"What I notice, being my first time in a hospital, is the constant queue of things which seem to happen to the patients: quinine, thermometers, nurses and doctors and seldom food. I feel I could eat like the proverbial horse but the hospital view seems to be that it is sufficient merely to keep the spark of life within the sickly frame. It all starts at 6 a.m. with a nurse taking a temperature, pulse, giving bicarbonate of soda, assumed to settle the stomach, quinine (ten grains) to settle everything and a cup of tea to prevent any hard feelings.*
>
> *After the ablutions comes breakfast. Food in bed is a torture but they do their best to interest the patient with a sample, alas that is all there is.*
>
> *In the middle of the morning the doctor, with a nurse or an orderly, would make their rounds. The good doctor asked was I still depressed? My only cause for depression was having to stay in the hospital! He asks about any aches and pains and even tells me where they should be found in reply to my saying I have none. He was right, as the aches and pains did eventually arrive."*

It is difficult to explain malaria to anyone who has not experienced or seen it first hand. It is often said that there are 'frequent bouts of flu-like chills', but when one lists the symptoms; fevers, aches and chills, it sounds like the common cold; but it is far, far from anything like that. The best description I have found was by a Polish man in West Africa, Ryszard Kapuscinski who wrote in his memoir, *The Shadow of the Sun*:

"The first signal of an imminent malaria attack is a feeling of anxiety, which comes on suddenly and for no clear reason. Something has happened to you, something bad... Everything is irritating. First and foremost, the light; you hate the light. And others are irritating—their loud voices, their revolting smell, their rough touch.

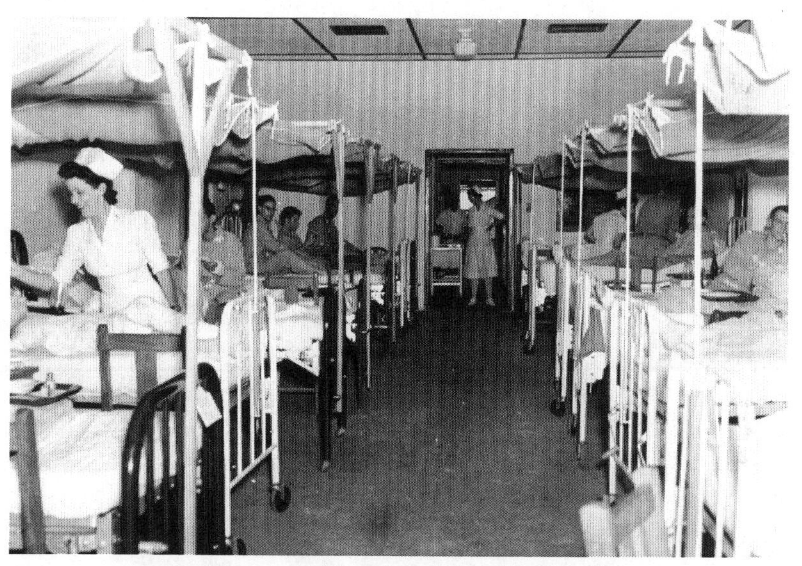

British General Hospital, North Africa c.1943.
The patient on the far right is smoking a cigarette in bed.

But you don't have a lot of time for these repugnancies and loathings, for the attack arrives quickly, sometimes quite abruptly, with few preliminaries. It is a sudden, violent onset of cold. A polar, arctic cold. Someone has taken you, naked, toasted in the hellish heat of the Sahel and the Sahara, and thrown you straight into the icy highlands of Greenland or Spitsbergen, amid the snows, winds, and blizzards. What a shock! You feel the cold in a split second, a terrifying, piercing, ghastly cold. You begin to tremble, to quake, to thrash about. You immediately recognise, however, that this is not a trembling you are familiar with from earlier experiences—say, when you caught cold one winter in a frost; these tremors and convulsions tossing you around are of a kind that at any moment now will tear you to shreds. Trying to save yourself, you begin to beg for help.

What can bring relief? The only thing that really helps is if someone covers you. But not simply throws a blanket or quilt over you. This thing you are being covered with must crush you with its weight, squeeze you, flatten you. You dream of being pulverised. You desperately long for a steamroller to pass over you…

…A man right after a strong attack of malaria is a human rag. He lies in a puddle of sweat, he is still feverish, and he can move neither hand nor foot. Everything hurts; he is dizzy and nauseous. He is exhausted, weak, limp. Carried by someone else, he gives the impression of having no bones or muscles. And many days must pass before he can get up on his feet again."

On the second day, I experienced my first of many attacks, and I don't think the Polish words exaggerate it at all! After the attack, I was a wreck and fell asleep so the Padre, who was very good, came and left two books by my bed, being careful not to wake me. I must have been in a deep sleep—which is unusual to achieve in a busy hospital—and I never heard him come in, or any of the others bustling about and clanging their large silver trays, or someone coming past my bed, but I never saw the books, which apparently someone took while I slept.

I had been taking liquid Quinine for a few days, which was pretty terrible stuff. The side effects really were bad, and I think I must have been pretty sick, because I didn't even wish that my quinine doses had come in the form of tonic water with a shot of Gin. We had often joked about Gin and Tonic being a 'purely medicinal' drink, but when it came to actually having malaria, I don't think I would have handled it too well. In any event, a fellow inmate and I calculated that if we had gotten the necessary dose of quinine in Gin and Tonic form, the malaria would have been the least of our problems—we would have needed about 15 gallons of the mix per day!

Eventually I was allowed to sit in a chair for a while, and then even walk a few steps and finally the good news came that I could leave the hospital. The Padre came and collected me, and he and the driver together with myself sat in the front while another officer sat in the back seat. The Padre was speaking about health in Africa, and whatever it was he said gave rise to a comment from the man at the back. Followed by:

Padre: '*Ejaculations from the rear!*'

Driver: '*Do you get them there as well?*'

On the 24th February, two weeks after my initial fever, I was back in the mess hall when I journalled:

> *"After what I suppose was determined as the correct period for recovery, I under went the usual aircrew medical, which included the examination of the spleen. I was able to maintain the mercury for 100 seconds, which I thought should have convinced anyone that I was fit. But no, my 'spleen'—whatever that is—was 'enlarged'. So the board was put off for another few days."*

Finally on 27th February 1942, I was passed as 'fit' and back flying. One week later, after leaving Wadi Saidna with the next convoy to Egypt, the lead aircraft (the only one with a radio) received a message of an approaching sand storm. Accordingly we were diverted to a landing ground at Atbara, somewhere along the Nile in northeast Sudan. As we landed it was possible to see a wall of sand in the distance—fast approaching. Taxiing in, I damaged the tip of one blade of the propeller running into a heap of sand. Fortunately the Air Force had thoughtfully provided within the machines the canvas covers required, so we hurriedly covered our aircraft. We were taken to the local Army Officers' mess where we lodged and where we were given a meal, which was good—if not a little gritty. We then got the message there could be no progress today and so we returned to our aircraft to retrieve our baggage. En route, however, a most unfortunate incident occurred which I meticulously journalled on 5th March 1942:

> *"The truck was the usual service machine with metal seats lengthways on each side but being an open*

vehicle, the African sun made them unbearable to sit on, especially when wearing shorts. I was standing up so I got a good view of the road ahead and the ensuing events. There were between seven and eight of us onboard. Our route took us along a street with the usual high walled Egyptian houses on one side and on the other, a ditch, an open space and a railway line. There was no sign of life except for an old man walking along the side of the road. From a distance away the driver blew the horn, which was met with a lazy turn of the old man's head. He continued on his way and soon we were within a few yards of him when suddenly his lethargy left him and he turned, looked in our direction and dashed out directly in front of us. We hit him with the full weight of the vehicle. Seeing him in the road behind us it was obvious he was past all help. The driver went to find a telephone to summon someone. There was little any of the rest of us could do other than have a cigarette and wait. The day was hot, dry and oppressive as only the Sudan can be.

Strangely the event attracted no attention for a very long time, I suppose it was siesta time and they were all asleep. Eventually a police car and an ambulance arrived. The slim, tall Sudanese police officer walked around with an air of efficiency and with a tape he measured the wheel tracks. Then someone put the old man's hat over his face. Two white-aproned Sudanese men from the ambulance produced a stretcher but they did not use it; instead they stood in the shade and told jokes. Last week's malaria must have caught up with me, for all of a sudden I felt very tired and desperately longed to return to the mess where I could lay down.

We were told we must all go to the Police Station. Perhaps it will not be so hot there, I hoped. We got into another truck. At the police station we were conducted into a room with benches around the walls. On the wall was a map and two charts and an index of 'strength and establishment' of evidence. In the centre of the room was a desk at which sat the slim police officer, and on the desk in front of him were a few loose papers. He had a conversation on the telephone then looked at us and asked our names in turn. This seemed to give him some difficulty. He sent for a glass of water—not for me but for our driver.

He asked who saw it, 'Well, yes I did!' I said.

'What exactly did you see?'

'I saw the road and the old man walking by the side of the road. The horn sounded and the old man turned and then dashed into our path'.

The policeman wanted me to sign it. I looked and saw all these disjointed sentences. But I signed it anyway.

Everyone else had to make his statement too, which seemed to take hours. Then we were told we would have to make our statements on oath. That meant waiting for the Magistrate! I waited outside in the shade where there was a little girl who was taking two dishes of food to a prisoner. The prisoner seized the food in his manacled hands and threw it across the courtyard. He looked a strong man. The jailor put him back in his cell and closed the door saying he is mad. 'Lofty' Dafhorn, another pilot, walked past his cell and the man asked him to see that he got out, as he said his 'Mother will be upset without me'. Dafhorn talked to the jailer who said the man would be freed tomorrow.

Eventually the Magistrate arrived, doubtless having completed his siesta while we were hanging about. We took a dusty book, which I presumed was the Bible and the police officer recited our statements. The Magistrate recited some formula at an incredible speed, which he invited us to repeat. Our only hope was to look determined and articulate sounds as closely as possible resembling his 'I swear mutter mutter mutter truth mutter mutter mutter God'.

At last we thought we could go! But no! The brakes of the truck had to be tested. And anyway they seemed to want to lock up the driver! We tried to dissuade them from this while the brake expert was brought in. Soon a perspiring little man would arrive with some instrument in a leather case. Eleven hot, sweaty and tired men watch him fix his instrument to the truck and drive round the square of sand, now and then putting on the brakes, and he said, 'they are no good' but admitted that he didn't know how to drive the truck anyway! Our driver got in and demonstrated and every time he tried to brake the wheels locked and the vehicle stopped in a cloud of dust. Brakes were perfect! They accepted this and then we could go on our way.

We were driven to a local army camp where we were accommodated in tents and weathered the sand storm that duly arrived and buried us! But this did not protect my mind from seeing the image of the man laying on the dirt road as I tried to sleep."

It was said at the time, that it was quite likely to have been 'Snail Fever' or 'Bilharzia', the common names for Schistosomiasis, a disease that caused the man to seemingly acknowledge us, and then dash out in front of the truck.

I suppose it could well have been that, or alcohol or almost anything else, but the reason it was thought to have been Schistosomiasis, was its incredible prevalence in that area—only second to malaria in terms of death rates. Admittedly progression to neurological involvement was less common, but when it did happen, a person would often have a combination of impaired judgment, slow reflexes, difficulties coordinating movement, blindness and deafness—none of which would help a person to cross a road. So I suppose it was a plausible theory, but being in Africa, I am quite sure nobody ever did much to determine the exact cause.

Schistosomiasis is a parasitic disease caused by a blood fluke or flatworm, whose larvae use freshwater snails as host. The thought of it makes me cringe. The snails then produce large numbers of larvae which are capable of penetrating human

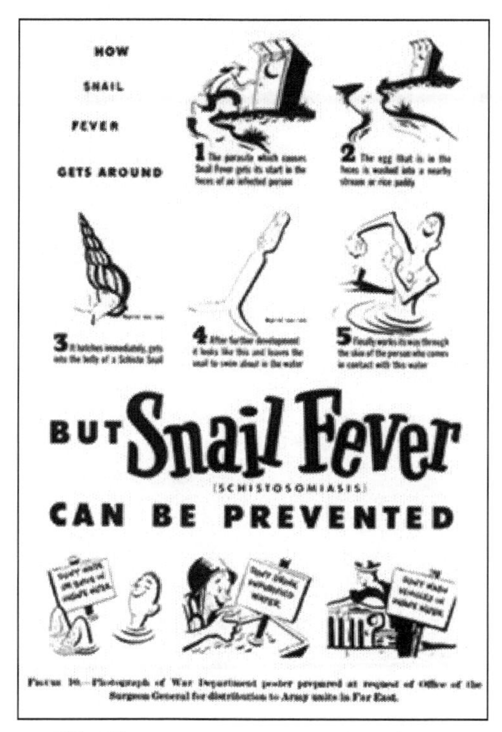

War Department poster promoting a preventative approach to Schistosomiasis, 'Snail Fever'.

skin. These are often a problem when one is bathing or walking in bare feet along a river. Once the larvae have penetrated the skin, the flatworms live in the veins, bladder and large intestine of the person, until they reach maturity. They then mate and produce thousands of eggs that damage whatever organs they settle in, including, sometimes, the brain. The eggs are eventually discharged through urine and faeces into polluted water sources and so the cycle begins again, with the next poor innocent person. The disease was hideous and to me it sounded nastier than malaria, but worst of all, it was so common.

It is so prevalent that seventy years after the end of WWII, the World Health Organisation states that Schistosomiasis still claims 750,000 lives per year in sub-saharan Africa alone. Not surprisingly, from 1940-1942 the Germans began experimenting with methods of preventing the infection from affecting their African based troops. They were relatively successful. Meanwhile, in 1942, the British unfortunately had their worst experience with the disease. The West African Force saw 432 British and 1,279 African troops develop the disease after an exposure in a lagoon in Nigeria. After this it became something that we were all highly aware of and regularly briefed on. The RAF was very quick to act on any potential exposures; if ever you fell in the

Along the Nile in Northern Sudan, a man drives two oxen, who in turn cause the large paddle wheel to spin and move the water to a higher elevation where it can then flow along a channel to the crops.

river, they would suddenly rush you off to hospital to be washed down, with some sort of alcohol solution. We didn't take any chances, no matter how tempting a swim may have been on some of the hot African days, the thought of these little critters was enough to keep us from so much as dipping our toes.

While we were still in Northern Sudan, waiting for the sand storm to pass, I went down to the River Nile, to see if anything interesting was happening. I was amazed to see a man operating an irrigation system. How they didn't get dizzy I don't know, but the man drove the oxen around a circular wheel, which was made of wood. This in turn spun a shaft, which caused a large wheel, to spin and bring water up in heavy, wooden buckets. The water would then be poured into a channel, and flow from its elevated level, around and behind the irrigation system, towards the crops. It was an incredible design, so rough and basic on the one hand, and yet so intricate and complex. I was very impressed with it.

Two days after the accident, I wrote:

"We received reports of the sand storm having passed and so we decided to carry on with our journey.

We were helped to remove the sand from our machines by a couple of mechanics. Having removed the covers I tried to remove the sand from my engine with cotton waste. I got the mechanic to produce a hacksaw and we cut off the damaged propeller tip and using this as a template trimmed the other two blades to match. My machine seemed to have more sand than the others. It was everywhere! Everyone else got their engines started and run up and I had no signs of life in mine. I watched the rest of the formation take off and resumed the inspection. Presently a fitter arrived and suggested I go back to the mess in time for breakfast while he removed all the plugs and checked the leads. That seemed a good idea!

When I returned to the airstrip, the fitter was happy to tell me that he thought it would start. After a couple of attempts it came to life and gave the full boost and revolutions. There was a little vibration at full revs but nothing serious. I was able to take off and fly following the Nile until it bent westwards and then follow the railway line to Wadi Halfa. I cruised at a little less than the normal boost to prevent any vibration. When I landed, I found the other four—three of them with damaged propellers!"

8th March 1942

"At last we landed at Cairo and I lost my hat which blew out of the cockpit on landing. I was not attached to it!"

It was good to have landed and to be back on my feet, after a long battle with malaria, a horrible accident and two back-to-back trips across the Takoradi route, I was looking forward to getting to the wonderful mess we had in Cairo—the houseboat.

Chapter Eighteen
Further East

IN EGYPT, WE HAD A squadron headquarters known as Kilo 17, Fayoum Road, where various detachments cooperated in the form of repair, maintenance and salvage units. I was there for a few days, until on 16th March 1942 it was time to fly again:

> *"At Fayoum Road I air tested a Blenheim. This was a shuddery short 10 minute flight in which it was apparent why I was to take it to a maintenance unit at Habbaniyah in Iraq. After a little fiddling by the ground crew, the second flight was a little more encouraging and it lasted 30 minutes which was enough to convince me that we could get there. I was to take a passenger with me to Habbaniyah and he must have been keen because he accompanied me on the test flights!"*

The next day:

> *"We set off for Aquir, somewhere in Palestine where we had lunch and then to Habbaniyah, which is close to Baghdad in Iraq. Palestine looked incredibly green and fertile as we passed over it. Iraq seemed to be much more desert and when we arrived at Baghdad it was very cold. A local told me it had snowed yesterday!*
>
> *The RAF station is very large and I understand it cost a fortune to build. It contains cinemas, churches, swimming pool and all the amenities one could wish for. During a recent uprising some hostile Iraqis had obtained a field gun and were able to shell the*

airfield from the nearby hills and did some damage to some aircraft. The mess hall is very well equipped with every comfort including a gramophone with a record, which someone played interminably. It was a song by a lady who repeated the refrain 'Please do it again, I may say No, no, but please do it again' with appropriate sighs and groans. Someone had managed to cut the record so that the noises were repeated ad infinitum. Another record which the locals had almost worn out included the words '...the concert was over in Carnegie Hall, the maestro took bow after bow...' the title of which I never knew, but I believe the singer was Ella Fitzgerald."

I was supposed to return to Cairo the next day but I was told the flight would be postponed because of a headwind. I went to look at the aircraft, which was going to take me back to Egypt, and noted:

"It is a very elderly machine, a biplane with a lot of flapping canvas. I can hardly believe such a machine exists outside a museum. They told me that until recently it had operated in Ethiopia, and to prove it, I was shown the special bomb sight! This comprises of a triangular piece of wood and wire attached to the outside of the pilot's side of the cockpit. [The pilot] has to put his head out into the slipstream to operate it! This machine is so slow it is the only one which can fly up the narrow canyons where the enemy is situated, so they said. My confidence was not increased!"

I returned to the mess hall and the gramophone, where the 'Do it Again song' was still in play. I think, the lady now singing was Judy Garland, but in any case, I had heard quite enough of the same line! The next thing I noticed was that the few people in the anteroom suddenly and quietly left by the back exit. I should have realised there was a reason—which soon became apparent. Someone, it turned out to be the Adjutant, approached me from behind, saying, 'You fly Blenheims, don't you?' but it was more of a statement than a question. His message was that someone was needed to fly a machine to Singapore and I appeared to be the someone. It was explained to me that the particular machine was to have been flown by a Polish Squadron Leader Kurdzel, but he was now unavailable.

I didn't even know where Singapore was, except that I would head south towards India and then turn left. I made my way to the hangar where the Blenheim was parked. By asking a few questions of the ground crew, I was able to piece together the story, which somehow the Adjutant had forgotten to mention. It happened that during the 'troubles' and the Iraqi uprising, a field gun had done more damage than I had imagined. Several aircraft had been hit, some, very badly.

The news of the Japanese invasion of Malaya back on 8th December 1941, had inspired the ground crews. The men had reconstructed this machine using salvaged parts from three victims of the Fayoum Road shelling. They did it in their spare time. It was to be their gift to Singapore, because they felt that Singapore needed all the support and encouragement it could get. At this stage I had not yet heard the news that Japan had invaded Singapore eleven days ago, on 8th February 1942. I wasn't told a lot about this mission at all, but it would have been nice if they mentioned that I'd be landing in a Japanese occupied air field, if I made it that far! A Polish officer had agreed to take it, but he required several repairs to be made first, and finally a coat of paint, so that 'At least it looked serviceable', and in the meantime, the said officer had apparently had a stroke and was now in hospital. Of course that sounded very nasty and serious, but I wondered if being in hospital and away from this blessed machine made him more likely of a better outcome than me.

My first flight in 'the Gift to Singapore' was very short as the sliding canopy slid right off and went on a cross country flight by itself. Then on the second, the starboard engine developed an unhealthy drop in oil pressure.

I enquired about the crew and it seemed they had all found other jobs. I asked if there was, at least, a navigator and a radio operator I could take. For one reason or another no crew could be found. However, the Adjutant later told me he had found a volunteer radio operator, Flight Sergeant Rennie. He seemed a pleasant fellow but pointed out he would be no more than a passenger, as the radio did not actually work!

The Adjutant was determined to get rid of this machine and me, presumably, before I came to my senses, and said 'No'. He rushed in with the news that an Australian Squadron was about to refuel here on its way to the Far East. 'If you leave now, you can tag along with them' he cheerfully announced. The squadron was passing through in three flights, the first having just landed. I made myself known to Squadron Leader Monroe and he told me to fly on the starboard wing of the formation. I did, but that flight was short lived, as, on takeoff, the oil pressure dropped below 35 lb so I had to return with one engine out.

All was not lost. I was told they could easily fix a 'new' pump (nicked off some other aircraft, of course). I could then join the other flight, which was about to arrive and go on the next day. I duly made myself known to the leader, Flight Lieutenant King. He was less communicative than Monroe. This time, however, we did get into the air, although it was a bit of a struggle as their aircraft were fitted with engines which had + 9 lb/sq.in. boost which they insisted on using for takeoff. My horsepower was much less and I was not happy to overload anything on my 'home made' machine. They were terrible 'formation' pilots and I could seldom see more than one of them at a time from my position, and sometimes none, but as they all had working radios and operators they were not worried. To make matters worse the formation leader liked to map read in his own particular way. I had no radio, but my operator sat beside me and fed me sandwiches, coffee and some Palestinian wine! Socially, I was better off, I suppose, but physically I think I would have been better flying the clapped out old biplane in the other direction—back to Egypt!

We stopped the night at Bahrain, a large area of sand on an island in the Persian Gulf. Accommodation was good, given that it was the B.O.A.C Rest House.

On 26th March, 1942 we left for Sharja, Oman. The weather was not good and most of the way we flew along the coast at a few hundred feet. After a somewhat sandy lunch, we took off and climbed to get over the mountains on the peninsula then down again to a few hundred feet, along the coast of Iran. As the visibility got worse, we flew lower. It was difficult to follow King's formation as he twisted and turned along the coastline. I seldom saw any machine other than my immediate neighbour, until apparently King suddenly came to a headland—evident to me when he suddenly did a steep 180 degree turn. I had to do a lot of evasive action to avoid him whilst trying to keep him in view. The whole formation was apparently in reverse! I followed, occasionally losing them in the mist along the shoreline. The formation, or what I could see of it, suddenly turned inland. My first reaction was that we must have reached Karachi, but that was impossible as we were going the wrong way, and the landscape definitely did not match the map!

Almost immediately, we were surrounded by high rocky hills, which loomed up out of the mist, some of them higher than us. This looked very dangerous to me. Way up front, a big flash—probably the explosion of the fuel tanks on impact—convinced me this was not a healthy place to fly. I could see cloud enshrouded hills on each side. We had snaked our way through a sort of valley, and those of us who had avoided the rocky hills, had done so by trying to fly under the cloud and winding around the lower bases—but it was all a game of chance. The mist made any unusual shaped or tall rock formations and cliffs invisible until the last seconds. I felt we should reverse as best we could. The visibility was down to almost nothing, and we had already lost the crew of at least one plane, and I wasn't sure how many others.

Now, separated from our convoy, I did the steepest turn I had ever managed in a Blenheim and flew back to the coast and continued eastwards. I remained at about 50 feet, any higher and the ground was difficult to see. It was about a half an hour after our 'estimated time of arrival' at the mouth of the river at Karachi. I could recognise nothing positive until eventually I saw a possible river estuary; I thought we may have gone past where we should have turned inland to the airfield. I asked my wireless operator if he was able to get any sort of bearing, but of course,

Rennie in front of the Blenheim after it had taken a nose dive into the sand of Karachi Beach, Pakistan.

that was impossible, from so low and after we had gone so far off course.

I decided that we would have to land while there was some light and ask where we were; a desperate measure in any event. Rennie agreed it was worth the risk to land with the wheels down, as it looked like firm sand and it was the best chance of ensuring the aircraft would be able to take off again. Using a pile of rocks as a marker I did a circuit, three legs of which were blind—in low cloud— accomplished using my watch. My course keeping must have been good because our base leg brought us exactly over the pile of rocks! The landing was perfect, except for a problem when we had almost come to a standstill. We struck a patch of soft sand and the aircraft suddenly put its nose into the ground. Rennie was unhurt except for a scratch on his hand. I had slid forward and past the end of my seat as the nose dug into the soft sand; when the tail bounced back down, the harness pulled me back tight, where the bottom of my spine connected with the end of my metal seat pan—very painfully. Against all my natural urges, and the pain that reverberated throughout my body, I had to get out. I didn't have time to feel the pain.

The nose of the Blenheim looked like I felt, with broken and dented parts everywhere; she was damaged beyond repair. We had no idea where we were, but wherever it was, it was obvious that we had been observed, because approaching along the beach was a group of native people. Not knowing how welcome we would be, I left Rennie seated in the machine gun turret—although it was pointing the wrong way and into the air by this stage. I staggered forward, rather painfully, to meet them. I was relieved when I got close and was greeted by a friendly 'Salaam'. We shook hands but we were unable to converse except by hand signs, and some very broken English. They indicated that we were welcome to go to their village but we declined, which seemed to worry them and later we found out why. They were able to tell us that we had landed in Pakistan, so we asked them to take a message to Karachi, which they seemed to understand, and promptly did.

They left us, and we prepared for the night. My back injury had now become very painful; so much so, that it meant that I could not make myself comfortable—neither laying in the sand nor in a variety of other positions in the cockpit. Eventually, I tried to use a parachute slung from the tail plane to the wing like a sort of hammock. That was not a success either and in the end I slept across the tail plane, the surface of which was hard, but with some parachute padding was tolerable.

In the morning a man who looked like the oldest inhabitant, brought us some milk, in a very good looking silver bowl. He made a fire and brewed the tea. We gave him some small coins and indicated we would like some matches. We intended to light a signal fire. He soon returned with matches, eggs and materials to prepare food for us.

A Hudson and a Blenheim flew over us and fortunately they saw us. The Hudson soon circled back and dropped some brown spotted bananas, which were not improved by the landing! Attached to the bananas was a note, asking if the aircraft was okay and telling us that a minesweeper was on its way, due about

Rennie posing between two of the armed guards from the village.

7.30 pm. They had advised the minesweeper of our location. Of course we had no way to reply.

We waited for the rescue vessel. When it failed to arrive we decided, as it may be looking for us, that now would be the time to use the distress signal from the aircraft's dinghy. In the course of looking for that, we discovered that we did, in fact, have a few machine guns on board with some ammunition. The instructions on the distress signal said to pull the tab. That took a lot of energy and contortions and it did occur to me that holding it between one`s knees whilst pulling the tab could result in an unfortunate injury! The flare successfully went off. Having no reply, we settled down for another night on the beach, but this time, inside the aircraft to avoid the cold. Two of the village elders remained with us, and they were armed. This alarmed us and later, we learned why. Rennie eventually slept on the beach wrapped in a parachute, while I tried to make myself comfortable on the pilot's seat in a semi-vertical position: very uncomfortable. Then I tried the tail plane again, which was hard but probably the right thing for such an injury.

As the first light of day came, I woke up and scanned the horizon. Imagine my joy when I saw, in the distance, a dull grey shape shrouded in mist. I saw it turn as if coming towards us, and decided to set off our remaining signal flare but there was no reply. The ship was going away from us. We felt pretty down,

Cyril Johnson, in his pyjamas, sleeping on the hard surface of the tail plane, after he had severely injured his lower back.

so we made some tea, in total silence. As the morning wore on the sun became a merciless fire.

That afternoon, while we continued waiting, and hoping, I happened to see a man appear over the sandhill. He carried a long rifle, the type used in that part of the world for many years. He obviously saw us and his intentions did not appear to be friendly as he unslung his rifle. Fortunately our guards, who apparently had not been noticed, saw him and a brief exchange ended in his leaving the scene much to our relief.

We draped the parachutes over the wing again with our local guardians under the other wing. We continued our endless games of cards, occasionally looking out to sea. A couple of days later, the minesweeper again appeared. This time we had no distress signals left so we took the cartridges from .303 machine gun that was in the turret and soaked the cartridges in petrol, before setting them alight. We soon had a signal fire. They replied, and as they came closer, we stood waiting in the sun. A smaller boat was lowered, bringing with it two men who were to attend to the salvage of this blessed Blenheim. We were soon aboard the minesweeper.

Once aboard, we traded some of our ammunition and a couple of spare machine guns for a crate of beer! After which we proceeded along the coast to where we picked up the crew of Blenheim number 4 who had walked back to the beach. I heard that one of the Australians had landed and somehow got hold of a horse and rode it from miles inland to the beach. I am not sure that anybody else was ever found.

Shortly after, our minesweeper arrived at Karachi, Pakistan. It was interesting to notice that we had almost gotten

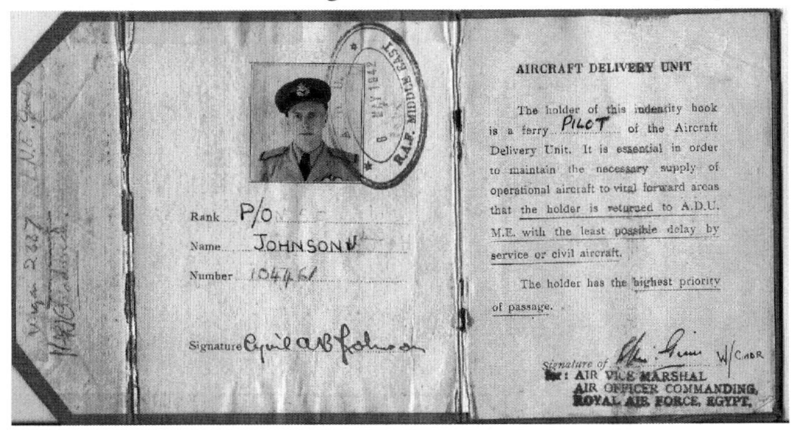
R.A.F. MIDDLE EAST

Rank P/O

Name JOHNSON

Number 104461

Signature Cyril A.B. Johnson

AIRCRAFT DELIVERY UNIT

The holder of this indentity book is a ferry PILOT of the Aircraft Delivery Unit. It is essential in order to maintain the necessary supply of operational aircraft to vital forward areas that the holder is returned to A.D.U. M.E. with the least possible delay by service or civil aircraft.

The holder has the highest priority of passage.

Signature of W/CMDR
for: AIR VICE MARSHAL
AIR OFFICER COMMANDING,
ROYAL AIR FORCE, EGYPT.

Cyril Johnson's Identity Card, stating that he has the highest priority of passage.

there before. The Station Commander asked if we needed to see the Medical Officer but we both felt that a good night's sleep was what we most needed—although my back was very painful. The next day the Station Commander showed me around the maintenance workshops and other facilities on the station. I was impressed! Rennie and I got a taxi and went into the city and sampled the night life of Karachi in a hotel with good beer.

The more we heard the world news for the previous few days, the more we realised just how lucky we were that we never made it to Singapore. It was the only thing that prevented us flying into the Japanese invasion, and I don't think we would have had a chance against their defences.

The next day, I used my identity card to get a ticket for Rennie and myself to fly back to Egypt. My identity card said

I was to have the highest priority of passage and it was with some amusement that Rennie and I displaced a General and his colleague. He was far from pleased that a mere Pilot Officer and Flight Sergeant had delayed his departure. On the flying boat we met a girl who had been working in the British Embassy in Iran. She soon struck up a great friendship with Rennie and me, especially Rennie, and she was soon offering him a sip from her bottle of 'Eau de Cologne', which, in fact, was the best Scotch. The Eau de Cologne bottle was needed as a disguise because alcohol was not allowed in Iran, presumably due to the Muslim influence. Working in a foreign Embassy, in addition to her powerful family connections, meant that she had access to the finest Scotch.

Rennie said he would like to see more of her in Egypt and I suggested he should make the most of the great friendship that had blossomed. Unfortunately, when we arrived in Cairo the lady was met by a group of very senior looking staff from the Embassy and she was, as it turned out, the daughter of a British Cabinet Minister. She hung about for the longest time making every excuse to delay, but poor Rennie was too overawed to do anything about it. I nudged him a few times, but he retreated in shyness; so ended a great romance.

Chapter Nineteen
Culture and Crisis

ONCE WE ARRIVED IN CAIRO, we had only a few days before we were to be flown back to Takoradi, to begin the ferrying process again. It would be one of the last times I would see Rennie, because as always, we were split up into various convoys—based largely on whoever was ready to go at the time—and we were constantly on the move, cycling through the landing strips and mess stations along the route. Rennie and I did however keep in touch through various letters and the occasional message left at a mess hall.

RAF men on Camels, in front of a Douglas C-47 Dakota, an aircraft often used to transport pilots from Egypt back to Takoradi.

It was an interesting culture we had here in West Africa; the RAF guys were all friends, because we had so much in common, and yet, I suppose because we saw each other in such spasmodic but unpredictable bursts, we rarely missed a particular friend. Either we would be busy catching up with some other pal whom we hadn't seen in a few weeks, or thinking we might see that particular mate fly up with the next convoy, or that he could already be installed in the bar at the next mess we would land at. We were like a deck of cards being constantly shuffled. There was always enough unpredictability that we never knew exactly who we would fly with, until the moment we all took off, and even that didn't guarantee who we would land with or eat with in the mess that night. There were often people dropping back due to aircraft maintenance issues, service delays, getting lost and turning back; or occasionally falling victim to illness, accidents or attacks and sometimes making a more permanent gap in the stream of pilots. It would often take a few weeks to notice if you hadn't seen such and such for a while, but usually the circle of communication was pretty thorough, and if you asked around you would always be able to find out from someone what was happening to a particular friend.

Before we returned to Takoradi, and in an attempt to distract myself from my lower back pain, which had not relented since the accident on the beach, I thought I'd enjoy a night on the town. The Opera House in Cairo had a claim to fame, being rather old and the setting of some famous operas; this time the production was Gilbert and Sullivan's *The Gondoliers* and the cast was largely drawn from the local British residents. I must have been in a sour mood by the time I saw the operetta as I wrote in my diary:

> *"...the music was second rate and the acting, even if good, was wasted on me by my not hearing sufficient of the words to form much of an idea as to what it was about."*

Once or twice I thought about seeing an M.O. about my lumbar and sacral spine, but the idea of being poked and prodded by someone who would probably tell me there was nothing they could do about it, did not appeal to me either. Besides I thought it was probably just a bit of bruising that would all go away in a few weeks; and I was content to try a glass or two of Scotch each evening, which was 'purely medicinal,' of course.

By 20th April 1942:
"I arrived back at Takoradi, in West Africa, to find piles of mail—mostly complaining about my not writing very often."

I rummaged through and put three envelopes to the top of the pile; the handwriting was, as always, a clear give away, and brought a smile to my face. I read the letter from mother first, and was put at ease knowing that she and my siblings were all doing as well as could be expected. Back home, the weather was starting to warm up. They had managed on the winter rations, but were looking forward to the promising produce of the garden. My sister's bee hives were doing as well as ever; at least the bees were oblivious to the war. I was delighted to hear that mother, who was always generous and gave everything away, had found that with sugar rationed, there was enough demand for her honey that she could ask for a bit of physical help in tending to the garden and maintenance of the home in exchange for her buckets of honey. She also reported that our cat—now almost seventeen years old—was doing well. Although she didn't write his name, I chuckled when I thought of mother standing on the porch calling Adolph to come and drink his milk. I considered what else would still be the same, and how much had likely changed, and suddenly I missed home and wondered how long it would be until I could see her again.

Choosing not to stay in any somber mode of reflection, I reached for the next envelope with Lily's unmistakable handwriting. There were two letters from her and, maybe more, deeper in the pile. True to her word, she had written me every week during the war; and she always kept me up to date on everything in the village, and all our childhood friends who had gone to their various posts for the war. She also wrote to most of our friends regularly and always seemed to be the central hub of information, letting everyone know who was doing well, or sometimes not so, and who had been posted somewhere with a new address. She told me that my friend Francis Ball was enjoying life as a navigator, and had been dispatched to somewhere in Rhodesia (now Zimbabwe).

Wigan was apparently, as I would expect, a lot quieter these days, and most of the businesses were not doing as well as they used to. Lily's mother owned a corner shop, and was rather good at business. Lilly informed me that her mother had started a system in which each customer could come in and buy what they needed, according to the standard issue of ration coupons, but if they could not afford to pay on that day, they could write their name and the amount owed into a little notebook. At the end of each fortnight when their pays came, they could settle the account for their groceries. The people in the village were all honest enough to honour their commitments, and it kept them loyal to her particular shop. There were often lengthy ration queues, and Lily commented that across the street at the greengrocer, she had seen my mother queued up for a few hours; apparently mother was trying to obtain some oranges—the one fruit we didn't grow—for Aunt Emma!

> On 24th April 1942 I wrote in my diary:
>
> *"I was promoted to the dizzy height of Formation Leader and set off in a Blenheim with a radio operator and seven Hurricanes following. Arrived for lunch at Ekejia, Nigeria, where I had never before landed. It was one very long runway cut out of the bush. We stayed over night in Kano. The next day we flew from Kano to Maidugeri—a nice short day. The pilots in my convoy were mostly Canadian or from the United States. They are quite good and keep close formation most of the time."*

A few days later, I was instructed to lead a convoy, and, almost as an after thought, asked if I would take a certain navigator who had not flown for 'some time'— the significance of this did not dawn on me before I had said, 'Yes'. I never got to know much about him as he was not particularly communicative, but at least my radio operator was an old hand and rather proficient. The normal course of the journey to Kano and Maiduguri was uneventful, and as always, the landscape from Maiduguri to El Geneina was very inhospitable. Mile after mile of nothingness

Bristol Blenheim.

and whatever landmarks there were changed completely when the season changed so one had to rely on the radio for this section or have brilliant dead reckoning.

All went well until three events occurred: first the radio operator advised that our radio had suddenly died, and then the new navigator panicked and went on strike. I think the reason for his panic, apart from having not flown for a long time, was that we were flying into an unforecast headwind. None of this would have been too disastrous had it not been for one of the Hurricanes coming alongside and waggling his wings, a sign that he had to land, presumably short of fuel. I dropped down over what looked a promising area for a forced landing, a grassy strip.

In my diary that day, I summarised the ending of this flight:

> *"Disaster. My observer is rather clueless and we were lost on the way to El Geneina, Sudan... I signed to the Hurricane to force land, before it ran out of petrol. I was ok because I had a bigger aircraft, a Blenheim, and could fly a bit further in the head wind, but the Hurricane had signalled that it couldn't go any further. I went on and landed at El Geneina. When I returned with help, I found five of the six Hurricanes belly landed."*

I later learnt that the other Hurricanes followed the lame duck and I landed at El Geneina all alone. The problem had been that we could not communicate, and so instead of just the one that was short of fuel needing to force land, the others had all followed him, assuming that they too could not make whatever distance was left. As it turned out, El Geneina was only a short distance further, and had I been able to tell him this, I think he would have been able to fly for the additional ten minutes or so, but he had no way of knowing how much further we had to go, and I didn't know exactly how low on fuel he was.

I doubt if my navigator ever flew again. I never found out why he gave up; presumably he relied on the radio more than his own judgment. A lot of damage was done to His Majesty's property that day, but surprisingly there was no loss of life and little in the way of injury except for a glycol burn on the legs of one of the pilots. All but one of the Hurricanes were so badly damaged that there was no way they could fly again.

On 28th April I wrote:

"The Khartoum court of enquiry—presided over by Willie Mahargh—found, after two days, that I was responsible as Captain for the navigational error, and that we should have turned back to Fort Lamy. He didn't have any other suggestions for what else I could have done in this situation, but said it was, 'Just one of those things'. There wasn't anything else that could be done, but as Captain I could see that I was to be held responsible and I accept this."

While I was waiting to hear the outcome, I can remember that I got a ride in a car into Khartoum city and there was a sand storm so I couldn't get back to Wadi Saidna for the verdict until the next morning. As it turned out, the enquiry never revealed anything, except that I was responsible. There was certainly a bit of a black mark against my name, but beyond that, I don't recall any significant consequences.

On 9th May 1942, I had completed my flight to Egypt and returned to Takoradi. I did my stint as a duty pilot for a few hours before being picked up by a passenger plane. I was not feeling very well this day, as I had another bout of malaria.

By 14th May, I wrote: *"I feel much better today but I can not say how long this will last"* before having a casual chat in the mess with the Air Force Minister, my friend the Padre.

Padre Swallow was a very sensible and realistic man, who the men found approachable and subsequently he often had many difficult problems to solve. One such problem was a

One of Cyril's mates, reading a letter in the Takoradi mess hall.

pilot who had recently received a letter from a chap in England, informing him that the said letter writer had run off with the pilot's wife, and made an offer to buy what had been his house. The pilot then went to the Padre to ask for advice. Not very pleasant news to receive, I would think, and I can't imagine what the best advice in the situation must have been. Another letter came to an airman from a woman informing him that she had given birth in February and asking would he please pay up. The letter went on to say, that the baby may or may not have been his, and that she was now engaged to another man; the Padre told the airman to deny the charge.

While I was chatting with the Padre, across the mess, one of my friends opened his mail. Before long, his face reflected the classic expression and feelings we had all experienced while reading one letter or another. In a boyish moment, overcome by the temptation of the camera I had sitting with me, I took a photo—to tease him about it later. He heard the noise of the shutter, and looked up rather surprised, but considering the mood he was in, he was a good sport about it!

With more time on my hands to reflect and my pensive mood not dispelled by my poorly timed joke, I left the mess and went to my room to write:

> *"It's been in the news recently that Tobruk was lost [a major port on Libya's Mediterranean coast near the Egyptian border] and that the Germans may even be advancing across the desert into Egypt. The news also informs us that although Malta continues being attacked, the USS Wasp delivered a large batch of spitfires to them, which should soon turn the tide of their air battles. However food and other war supplies remain critical in Malta. Meanwhile, in the Pacific, Japan is continuing strong.*
>
> *I believe what little we had in credit was the slight Russian advance in winter and the recent thousand bomber raids in Germany, which one supposes will wear the Germans out in time, but no one knows how long this may take. It occurs to me that the Germans are becoming desperate, like ourselves, and will put all their strength into the summer effort. They cannot afford to lose, everything is at stake—all their dreams of an empire will soon be dreams of an empire that has past.*
>
> *We too can only hope to crush the enemy. The Britain of Kipling is a thing of the past. If we fail we will have lost even that which we seem to have—so they tell us. The Americans and Russians even now are taking the war into their own hands and if they do, we shall all see the final battle between Capital and*

Labor, and I expect any compromise between the USA and Russia will be only in name."

Boys in the Mess at Takoradi, playing cards while drinking Club Lager and having a smoke.

In Takoradi we had a local newspaper called *The Pilot*, which was a civilian newspaper delivered to the town, and of course the RAF base. It was one of several African owned newspapers in the region—which focused on improving literacy and providing news. In this way it differed from the more established British owned newspaper companies, which seemed to have the primary objective of promoting and fostering colonialism, and of course making sales. *The Pilot* was a good read. It wasn't the height of journalism, but the local Africans worked very hard to produce it; I don't think many of them had an education standard much past primary school, but I appreciated their hard work and relatively unbiased reports.

There were also a lot of locals employed by the newspaper who had previously worked on trade ships. These men and women had been exposed to newspapers in various parts of the world, and they produced their own newspaper, which was a combination of all sorts of styles and layouts, taking ideas from all around the world. It was very enjoyable. The supporting technology for this sort of journalism must have been adequate because they were always up to date with the current conflicts and battles. They certainly had the radio, and they also had telecables and even facsimiles, so getting up to date news wasn't too difficult for them. Of course it all had to be censored, being independently owned didn't get them around this and although they were usually accurate news articles, there was the occasional bit of British propaganda in it too.

Life in Takoradi, during WWII.

Chapter Twenty

Malaria, Again and Again

IT STARTED IN TAKORADI, WHERE we had to wait several days on account of the weather. Despite being newly assembled, our aircraft were somewhat worse for wear, and I expect they may have already done their share of work before being shipped out here. We set off on the Takoradi route again, and for various reasons, needed to make night stops at every staging post, but we kept promising ourselves that the next night we would be in Maiduguri, which was a much nicer place to stay. We made it to Maiduguri, northern Nigeria, after a few days but having already lost time, we pressed on to Fort Lamy.

Bill Eily was leading for the first time with only a few hours flying experience on twin engine aircraft. Bill had been

Period postcard from Fort Lamy.

steering an erratic course, as he would admit, but I was at a loss to understand his sudden 180 degree turn until we got back to Fort Lamy. It seemed that his exhaust ring had disintegrated, and thus there followed a reshuffling of aircraft with Bill at the back of the formation. We took off again, and this time Bill had to fly a Bisley aircraft for the first time (a ground attack version of the Blenheim that was subsequently renamed the Blenheim Mk V). The flaps flew back on takeoff. We managed to get back to about the same point as we had the previous time when bad weather set in, and so, he did another 180 degree turn and headed back to Fort Lamy. We thought this was the best idea and all followed, but on this landing, my brakes failed. I was stuck at Fort Lamy until they could either be repaired or replaced.

29th June 1942 I wrote in my diary:

"So here we are in a very dirty little hotel—used as the Mess for the French Foreign Legion. Africa seems to take a keen delight in making me feel ill. Having spent a few days at this most desolate outpost, I awoke one day with a most acute headache and very stiff joints. My back—which has constantly caused me pain since crashing on the beach—and neck now made me want to bow low. A rather pleasant French doctor came to attend to me and told me this was malaria—again. He attended me regularly. I recall laying still with all over aches and pains and could not get up for two days. I was unable to eat. The very thought of food revolted me. I drank a lot of water and herb tea which the doctor gave me. I felt as though my head would never be the same again. The days passed without my counting them."

The French doctor in Fort Lamy was a strange man. His wife had moved out from Europe to stay with him, in this remote little town, just south of Lake Chad. When the doctor wasn't repairing the damage inflicted on airmen by mosquitoes or other African hazards, the husband and wife team would 'borrow' a Blenheim and fly further up north to drop bombs on the French—as their self directed contribution towards a small internal conflict, which was nothing to do with the war! The French were a bit strange: Fort Lamy was a bit strange too.

Three RAF men from Cyril's convoy, armed with rifles, watch carefully for crocodiles as they walk along the river.

Two things I recall distinctly about Fort Lamy. Firstly, when we were unfortunate enough to need to spend the night, the French would always insist we had wine with our water, or water with our wine, or just plain wine, but never plain water. It wasn't until I saw where the water came from that I understood why. I was going for a walk when I noticed two of the staff come down towards the riverbank; one carried a tray on which several table jugs sat, the other stood further back watching keenly with a rifle. The first man cautiously made his way to the bank and filled each craft, holding it up to the sun, checked for any larger floaty bits and then occasionally he would empty a jug and fill it again. Once they were all full, they left quickly and headed for our mess hall. I hoped they used a boiler!

Secondly, I clearly remember needing to be very careful each time we landed at Fort Lamy. The runway was very short, and at its end, was a river, and along its banks there would almost always be a few crocodiles sunbaking. We would literally taxi to the end of the runway and put the brakes on hard, promptly turning our aircraft around to taxi back. If we looked out the window, there would nearly always be a few crocodiles, just lazing about on the bank almost parallel with our turning circle. I had been incredibly fortunate when my brakes had failed because the wind was such that I hadn't needed the whole distance of the runway!

After a few days of French hospitality, and more river water than I care to imagine, my malaria improved to the extent that I was able to get up for meals and so I was flown back to Maiduguri as a passenger. Here, the Medical Officer I had seen previously, Doctor Scott, sent me to bed again with a course of Quinine. I felt quite tired but each passing day made me feel a bit better.

17 July 1942 I had recovered sufficiently enough to write:

"I am now reasonably well, except for the odd 'faint feeling' now and again. Sometimes, I felt there was a blanket over my head trying to smother me—just as it was when I was 12 years old and took two months off school with a strange nervous system complaint.

The insects out here are strange. They are many and varied and I have noticed that the rain seems to have awoken them from their hibernation. They descend on men in large formations. The only defence seems to be a mosquito net but even so, often insects get in on the wrong side of the net and indulge themselves at one's expense. One man went to bed with a scorpion—hardly a sexual perversion, as this beast is apt to turn spiteful. Moral: look before you sleep.

The African housefly—the same as his brothers in England—is the most common pestilent insect and is arguably the most frequent cause of disruption because it has the striking ability to infect food and cause dysentery, known here as 'the squits'. The next on the list of unwanted pests is the mosquito: possibly the greater danger. Specifically, the anopheles mosquito is the most dangerous, because it not only carries malaria, but also yellow fever. It is hard to distinguish between one mosquito and the next, so one must avoid all to be sure. There are all manner of pests out here, including the little Nairobi flies that can leave an acid over one's skin, which burns and makes small blisters, like mustard gas."

25 July 1942,

"The other day, some sort of missionary from the Leper colony came down and gave us a talk on the customs and superstitions of the local Kanuri people of Nigeria and the Kanem Empire. I was very interested in what he had to say."

I went on to research more in the days that followed this talk, and spoke with the local people to learn more about their culture, and they in turn told me who else to speak to and so forth, such that I collected a lot of information from both the locals, in their broken English, and the missionaries. I spent hours going around and speaking to people about their customs, and trying to make sense of them or understand how they may have come about.

The diary entry for 25th July went on to say:

"These people are mostly Muslims and are a village dwelling tribe. They plant seeds before the rain comes, unlike most tribes who sensibly enough plant it afterwards. These people plant it beforehand, as they believe they have the ability to know when it is about to rain. As their meteorological attempts are inefficient, they usually have to plant it several times each season. The Friday or Sunday before planting, a service is usually held to ensure the success of the crop. When it becomes apparent that the rain is not as close as hoped, more prayers are said. There is a strange method of bringing forth the rains: a large calabash—a

Traditional Kanuri People, a cluster of eleven closely related tribes across five North African Nations, photo believed to be taken in Northern Nigeria, c.1943.

large native fruit with thick waterproof skin that can be dried and hollowed out as a receptacle—is filled with water and another similar one is placed on top as a lid. The man then beats this like a drum until he is quite incapable of beating any more. The idea is to take the calabash to the chief and pour the water over him. Twentieth century chiefs—not being what they once were—usually see the calabash coming and rush inside their huts and close the door, so the farmer has to be content with swishing the water under the door.

There is great significance placed on the use of scars to make people look beautiful and distinguished. I am told the raised facial scars are achieved by placing ash inside of the fresh wounds, which helps promote these permanent markings once they heal.

African Kanuri woman in Northern Nigeria, c.1943.

It is a custom that all the agricultural operations are commenced on a Wednesday—never a Friday or a Sunday. When hoeing time comes around, it is customary for the people to labour communally on each other's corn patch, and it is also usual for a prospective son-in-law to bring his pals and hoe the father's corn patch for him. Not so welcome to the father as one may think—for he has to supply the food for a crowd of healthy young appetites.

When the first yield of corn shows, the

farmer goes through the paddock breaking off a head of corn here and there, and this is said to keep away the locusts. It is usual for the people, who are somewhat nomadic, to change the site of their village every five to seven years, and some even more frequently. Anyone who has smelt a native village will sympathise. One chief was very unpopular because he changed his site every couple of years; it is said that he now has very few followers left. The plan for the new village is always the same: a large square is selected and the Chief's residence is facing East at the end of it. The rest of the village is built around the square, which is left clear for markets and public affairs.

Wells are sunk on the hit or miss principle. No divining is used. It often happens that after digging a few feet, the earth becomes hard and unworkable. To solve the difficulty, a large quantity of groundnuts are thrown down the well and burnt. After a few days the digging begins again. The theory is that after a few days the oil from the nuts seeps into the ground making it soft and pliable.

To set up a market requires several teachers from the mosque. In fact they use as many as possible. The market is laid with a quantity of cow manure (or any other manure if they don't have cows) at the four points of the compass and in the middle. Then the teachers stand with at least one on each point of the compass and read prayers from the Koran in unison. The idea is that this will help to spread the rumour that corn is being sold very cheaply and attract buyers from around the area.

The taxation system is very simple. Each of the villages gives one tenth of his crop to the village chief—who in turn gives one tenth of all this to the local district chief, who in turn gives one tenth to the larger district chief. The total of all the 'tenths' by this time has reached a fair size, so the ruler gets quite a lot.

If one desires the removal of someone whom one dislikes, the method is simple: take the sand from his footprint and take it to the Emir [a Muslim Chief], who, with a view to reward, will hold the dust in his hand and read from the Koran. It is said that the undesirable then packs up and leaves the district and is seen no more. It sometimes happens that one doubts the fidelity of a friend. The cure is drastic; a frog is selected and long sharp thorns are driven through it until it is dead. It is then pounded by a pestle and mortar and the resultant mess is then put into the friend's food. The false friend then swells up and usually dies.

The locals also have numerous lockets worn around their necks—to ward off all manner of ills. They are mostly made of pieces of the Koran and crushed roots, contained in little leather bags. There are lockets to guard against all diseases and all the dangers of daily work. The herdsman wears a locket so that if a thief comes to steal the cattle, he throws a locket over his shoulder to become invisible and thus the thief does not kill him. This does not save the cattle however. Some cattle also wear the same locket around the neck. This makes them invisible to preying wild animals. There is another locket that prevents one from being harmed by bullets. One man says he definitely knew of a case where someone was shot at and the victim was unharmed."

It was all rather interesting. Looking back, I suppose the use of lockets as protection wasn't such a silly idea, and I figured that the superstition must have originated from somewhere. I would later learn that the father of my future wife had been in the trenches in World War I, and he wore a locket watch around his neck. The enemy opened fire and a spray of bullets headed towards him; a piece of shrapnel went through his shirt, towards his heart. It pierced the glass of the locket, but did not penetrate any further. In this sense, the jewelry around his neck had literally saved his life, so I could imagine how an incident like this, or any number of very practical logical events could have easily been attributed to something within their understandings. I could see how superstitions could so easily be formed in a tight community of oral traditions, and it was very interesting to think about how and why they had the beliefs that they did.

During the week that I had been talking with the village people, my body had been wracked with several severe bouts of malaria. At the time I thought a lot of what I learnt about this village was peculiar. However, I have since read up on these people and discovered that anthropologists have documented similar things, so my outlandish notes were not merely the result of my malaria! In any case, I mustn't have been making a great recovery, because my next diary entry, 2nd August, reads:

> *"After meeting all the local people at Maiduguri, I was taken as a passenger, up to Cairo and then onto Heliopolis, 'the sun city'. I was to report to the Heliopolis Medical board. The medical officer didn't seem to think much of me—a chap called Doctor Rumble. He took samples of my blood and I promptly passed out. He then returned me to the President of the Medical Board, who had been out for his afternoon siesta. The clerk didn't think he would be back at all that day, so I was told to call again tomorrow."*

All must not have gone well with the blood test results, for I wrote on the 7th August 1942:

> *"I am very tired and it is a long way to the hospital. Much has to be done on foot, and when one doesn't know the whereabouts of others, this is very tiring, and hardly good exercise for a sick and confused man on a hot day. So I returned to the houseboat, on the River Nile, where I had been staying, and having had nothing to eat all day—forthwith had a good meal."*

I was finally admitted into the hospital where I spent a week recovering from the malaria; and after being released, I went to Alexandria because I had to take a few days sick leave. The journey by train was without incident. Of course there was more humanity than could be comfortably carried—passengers even sat on the roof. That same evening I went into the town and, while having a drink at a place called the Monseigneur, I bumped into Reg Massa. Reg, his wife Helen and their young son, Paul, still lived in Egypt and it happened that they were dining in that very hotel.

13 Aug 1942:

"Douglas, the flight lieutenant in charge of the house, had organised a dance. Such dances were held almost every week. So tonight numerous lady guests came, mostly English and of every shape and size. I am not fond of dances, and my social inability.

Luckily, I was able to stand in the corner as a spectator and pay my contribution to the upkeep of an excellent bar. Two other men who were of a like nature, and opposed dancing joined me. We talked and drank until we were observed by an Irish girl, who lived in a flat above Reg and Helen. Her friends soon came across to us with the intention of breaking up our party. They succeed in removing one of us into the impossible melee but we two remained quite firm and were spared that horror.

I went into the dining room and found myself the last customer and no places available. However, after much bellyaching, I was able to find a place at the end of the table nearest the food. The conversation was difficult with all men trying to avoid certain words, that should not be said in front of a lady, but which they had grown well accustomed to using.

One unfortunate man staggered in to speak to his friend, but was too well oiled to be able to cope socially. Just at that moment, one of those curious silences overtook the company as he spoke, and in a very clear voice he announced, 'I am pissed', upon which there was a very noisy clutter of cutlery and the offender made an unsteady egress backwards and fell backwards onto the floor.

He was promptly removed by the hall porter and gave no more trouble.

After our meal we all adjourned, supposedly to dance again. My ally and I, however, chose a corner and repelled all attacks by these girls. This friend of mine would talk the hind legs off the proverbial donkey. Mercifully, this was the first time I had met him so I survived, as his stories were fresh to me. His education was a little imperfect, but it would seem that he had spent much of his time reading books, that he fails to understand—although he thinks he does—and he frequently uses large mispronounced words. Fate so worked that one of the women who tried to lure us out to enjoy ourselves was unable to comprehend words of more than one syllable. The conversation soon became a sermon—with the subject being the shallowness and ephemeral pleasures of this world. I gave up listening and heaved a sigh of relief when I could finally get to bed."

Egyptian style boats along the River Nile, with rented houseboats tied up at their moorings.

I spent most of the following morning just sleeping or laying down. The next day I went to Alexandria to meet Reg and Helen who were on their way to Sidi Bishr, on the north coast, to bathe. I spent 18 piasters on a pair of trunks (not a lot, even on my wage) and joined the party, we changed in the house that they had rented and walked across the road to the sea.

I wrote:

"14th August. Alexandria has several things to recommend it to the man on leave. There are cinemas and cabarets for those that enjoy such things as well as bathing in the sea or street cafés for the sober element. Alexandria has the largest promenade in the world. It snakes around numerous bays covered with bathing huts and sometimes, as is the case at Stanley Bay, in tiers one above the other. Bodies lie on the sand in such close proximity, that one has a considerable difficulty in walking between or over them.

On the landward side of the road are to be found buildings of unbelievable contrast. There are residences with numerous flat roofs and projections with rails and gangways. There are many hard bylies, small flats, where many not so well to do Egyptians go for a summer holiday. Even a small bathing hut with a floor 10x15 feet, eight miles out of town costs a lot. It is incredibly overcrowded—especially considering that this is wartime. The harbour contained mostly fishing craft. The other side of the harbor contains, amongst other things, the French Fleet—the crews of which are paid by the British Government who are always more generous to the French than their own men, this is likely because a few free French men told me that if the British government wasn't paying them, then they would join the other side—the Axis.

If you should sit down to rest and refresh your self at the café, you can see much of the commercial life of Egypt set out. Firstly, this cafe is of a French style, because Egypt historically had a great French influence. Make yourself comfortable and then the chap who cleans shoes will no doubt have reason to suppose that your shoes should receive his attention for the modest sum of a piaster or two. If he can bamboozle you into paying, and having got your foot on his little box, he goes to great pains to put loads of blacking on your shoe. He polishes it with such expert care that one can see one's own reflection mirrored in it. A tap with the back of the brush on his little box indicates he is finished and you need to put up the other shoe. Meanwhile, a seller comes along with nuts, trying to persuade you to purchase the minimum number of nuts for the maximum price. However he can be made to offer more nuts for the same price if he sees signs of you not having any at all. Having successfully obtained your nuts the man selling mangoes will find you fair game or, perhaps, the more impossible shellfish man. The haberdasher will suggest a smart line in check ties or sunglasses, or even filthy pictures. Sometimes the 'gili gili' magician will perform for you, and pull little chickens out of his sleeves.

Back at the base, Flight Lieutenant Douglas, as I said, organises things. Today he organised a bathing party on the beach. I felt rather tired but I still went. He invited some of the local 'Wrens', a nickname for the girls from the Women's Royal Navy Service. We duly bathed and ate tea and cakes. I felt very stiff today and my back ached like blazes. I then decided that the proper place on such an occasion was bed. Even there, alas, I found very little comfort and changing positions was almost impossible. Douglas gave me a couple of pills, as my temperature was also very high. I could not have eaten—I don't suppose I could have even eaten caviar!"

21 Aug 1942:

"Douglas sent for the local doctor who prescribed 'the local hospital'—the Sixty Fourth General Hospital.

I cannot say I went with alacrity. However, on arrival I was asked to stay—in fact they would not take no for an answer. They said I have sand fly fever."

Five days later:

"I left hospital and said goodbye to Reg and Helen and went back to Cairo. The train journey was much as before. I had not taken lunch, but there was a New Zealand nurse who decided to have tea with me; we got a bill for 27 piasters—it must have been good tea!

I returned to the Heliopolis hospital, hoping to be cleared, but they took my temperature and they said it was 'very high'. I was then admitted to the RAF hospital."

Over two weeks later, on 13th September;

"I am still in the RAF hospital in Egypt. Helen and Reg with their son Paul come to visit from time to time and bring a supply of cigarettes."

We were allowed to smoke in the hospital; in fact, this was so normal we didn't even question it. Smoking was not believed to be particularly unhealthy, rather, it was said to have some social benefits and many thought it was a useful means of keeping the mosquitoes away. There was certainly nothing in the advertisements to suggest they were unhealthy, and in any case, cigarettes were standard issue with our rations, although I was always sure to give them up for a few weeks at a time, just to ensure that I would be okay if the rationing was reduced or ceased.

Chapter Twenty One
Out of Africa

THE MEDICAL BOARD DECLARED THAT I could no longer stay in a tropical country and must be sent back to the UK. I was greatly saddened by this. However on that particular day, 20th September 1942, all I managed to record in my diary was the brief single line; "*Report for medical and was told I was to return to the UK!! I don't feel very pleased. I would have liked to have stayed for the winter*". Looking back I can see that I had not yet taken the time to really understand how much of an enormous blow it was and five months later, in February 1943, when I reflected on the significance of this day, more of my feelings emerged.

> *"On receiving the news that I was to go home, I fell into deep thoughtfulness tinged with no small amount of regret. On the whole I had enjoyed my time in Africa with its various sights and odors. All its familiar forces and characters—now it would be changed for I know not what. There was so much I had not seen and many experiences not yet enjoyed, and perhaps there would never again be the opportunity."*

On 21st September 1942, I made my way to the houseboat to say goodbye to what had been my home, whenever I was not in transit or back at the beginning of the Takoradi route. It was friendly, as always, and even if many of the people onboard were still newcomers and strangers to me, there were those who would great me with a smile and ask me to have a drink with them. Everyone was sociable and envied my 'good luck' as they chose to joke of the news that I was being sent out of Africa. The Adjutant quickly wrote off all my flying kit, together with the other odds and ends that I had lost, and put the paper on a tall pile already on his desk. The C.O. signed my logbook and a small trunk was put at my disposal to take my worldly goods to the transit camp. Someone arranged to have anything else that was still at Takoradi to be ferried up, but there wouldn't be a lot; I had been caught out enough times, with unexpected lengthy delays and hospital stays to always travel with the bulk of my belongings.

So as the morning of 22nd September dawned, I set out for a remote and particularly sandy camp. Here I discovered that I was to live in a tent until such time as arrangements had been made for me to move to Kasfareet Airfield. The mess was a converted packing case, or cases, and was quite hospitable even if the chairs were a little bug infested—a complaint common to most cane chairs and consequently it was always a mystery why they should be so extensively used in such a climate. My mess companions seemed to have come from all different units, and most of them were rather browned off with what they felt was a very long stay; some of them having been there more than three weeks.

There was a large Scottish man we called Jack, he was the first one to notice, perhaps because of his moustache or even because he had something to say on every occasion, and if no occasion arose, he would make one. He had been given a Distinguished Flying Cross for bringing back his Beaufort when he was badly wounded. Another interesting character was Flight Officer Barbary who had an arm in a plaster cast and who would be best described in his own words as, 'very pissy!' Later came a somewhat broad spoken northerner named Williamson and he kept us amused with some card tricks, and I taught him how to play chess, at which he soon became proficient.

Most of the day was spent playing patience, or double patience, at which Jack and myself wiled away many an hour. The day was not too strenuous even if it was a little boring, and our sole duty seemed to be to report to the Orderly Room every morning to sign the book and enquire about movements. In the evening there was the camp cinema to visit with a different film every night. It was owned by the Universal Shafto, who seemed to own every camp cinema in the Middle East. Attached to it, was a bar which sold gin, purporting to be Gordon's, but whose pedigree was held in some doubt. Beer, however, was not procurable except in the mess where a Canadian or Australian home brew was sold. On one occasion, someone said, "Suppose we go to the cinema and have a quiet night." We started with just a few sips in the bar and somehow remained in the bar, drinking a couple of Gordon's; no one was more surprised than us, when the film ended and the crowd turned out to see us still at the bar.

Maggie Carter, on the houseboat mess in Cairo.

We kept waiting at the camp, but it wasn't a bad lifestyle. It was a relatively simple matter to get into Cairo, as a lift could often be hitched all the way or at least to the station from which trains ran about every ten minutes. I made two such trips in to see my friend Helen, who on each occasion expressed surprise at my still lingering on in Egypt after the Postings Officer had initially said that I would only be waiting a day or two.

On 30th September 1942 Maggie came out to see me,

bringing with him my flying cushion, any loose belongings and some mail—about twenty letters. It was great to see him and he stayed for a drink so we could catch up on the latest news; who was where and what they were up to. He said we would have to keep in touch, and we made a plan to catch up after the war. I remember Maggie waving goodbye that afternoon, but sadly, this would be the last time I would see my friend.

In June 1944, the day after D-day, Maggie would be taking a load of ammunition to Caen, France to deliver it to the soldiers. There in mid air, he would be shot and the explosion of the plane and its cargo would kill him outright. The news of this I will never forget receiving; it came in a letter from one of the boys we had trained with.

Even after all the horrors we had experienced and were still yet to face, and the large scale deaths and battle tolls, we never did get used to such personal tragedies as the loss of an individual man, a friend.

Nearly two more weeks would pass, with me doing little other than waiting, trying to make myself useful by volunteering for duties and making short trips into town, and when this was not required, playing cards and supporting the bar. It was a quiet place and I was in desperate need of something to do, lest I begin to ponder, or search in vain to make sense of the war, something that was never helpful and which I greatly wished to avoid.

12/11/42

"Our Transatlantic Allies who have just arrived have suddenly overrun the mess. They are mostly air corps men and have been on a dry ship—upon their arrival, the first man had almost run towards the bar where they now all stood drinking, with their hats on, a custom I noticed peculiar to the Yanks, as apposed to our great southern Allies who were well mannered, if not also a little unruly at the bar."

The following day, the movement officer was running about looking for two people and being unable to find them he set his attention upon me. He asked if I would like to volunteer for a little job, although his definition of little was not specified. Given how boring life in the transit camp had been, I said that I would volunteer for anything.

> 13/11/42
>
> *"The Movements Officer explained that tomorrow I am to leave Egypt with 25 RAF men, mostly N.C.O.s to act as an escort; as an escort to what he didn't know. Furthermore he says that it will only mean that I go home a little ahead of the others—so maybe that will not be too bad. Apparently, the men are all very keen and say they will do anything to get on the way, so it seems we will get on quite well."*

The Movements Officer, who had advertised for the position, sent me down to the Suez Canal by means of an overcrowded train, with many passengers on the roof, and instructions to board a lighter at the docks. That was all the information he volunteered. It took a bit of drawing out, and a lot of prying questions for me to get an idea of what it was that I had signed up for, but slowly more details emerged; I was to meet and then lead twenty-five men from the Royal Air Force who were waiting on the docks for me, and then we were all to join a total of seventy others from the Royal Army who were already on board a ship. The Movements Officer seemed to think that that should suffice for instructions.

When I arrived at the docks and met the RAF men, a lighter ferried us across to an enormous ocean liner; a ship built by Cunard White Star company, called the *Mauretania*, built only four years earlier, in 1938. She was similar to most of the ocean going ships of her time; although not quite as big as the *Titanic*, but the height of a several storey building, 89 feet, with the same sort of funnel line. She looked solid, and she would want to, with the dangers that could come upon her from below the surface of the blue, or above, should some bomber pilot decide to drop his load.

As we were boarding the *Mauretania*, the RAF men duly informed me that I was to lead them, and then proceeded to give me a sort of briefing, seeing there had been some communication breakdown. My mission, now that I had no choice but to accept it, was to lead these 25 RAF blokes, and work with half a dozen designated British Army officers, who had all been set the same specific task. We were to guard five-hundred German Prisoners of War (POWs), who were being transported from Egypt to San Francisco, USA. I didn't know San Francisco was on the way to the UK!

RMS *Mauretania*, in her grey colour scheme, painted for war in 1939.

Apart from this designated handful, it became apparent that most of the seventy or so Army men who were already onboard were either there for purposes unspecified (certainly to me and possibly to them), or simply travelling on the ship in order to be transferred to their next destinations; they weren't interested in guarding any POWs. Also on board were a few people from various Allied military backgrounds, many of them clearly top brass sort of fellows, who we were to 'drop off on our way past' Australia and New Zealand. Following this, the plan was to 'call past Pearl Harbor, Hawaii' and then take the others across to North America, 'on our way home' to the UK. That certainly put any idea of inconvenient car-pooling clubs and school drop offs into perspective. The trip from Egypt to England was taking a detour—around the world!

14/11/42

"We inspected the prisoners quarters that were in the lower decks of the ship, all very neatly barbed wired off. Two of my Army companions are Newton and Rose. Rose informs me he is 'A/C' but I don't know why as Newton seems to have the most experience. Nonetheless, we spent most of the afternoon planning

and working out the best way to keep everyone safe and in order."

15/11/42

"Enter the ferries! All 500 POWs are going down to the lower decks, except for a Medical Officer called Feicer whom I guide up to a cabin on the promenade deck next to the hospital. To make room for him, I have moved from my cabin into that of Rose and Newton—and we seem to get on quite well. I get on with Newton especially, perhaps because he is a Yorkshire man. It seems that Newton and I will take it in turn to cruise around from time to time to wake up the sentries and see that the POWs do not escape. Tommy Rose will follow the O.C. troops Lt. Colonel Lewelyn around like a pet dog—with a notebook in hand."

16/11/42

"Set Sail! I am still rather sorry to leave Egypt. Perhaps one day I shall see it again."

21/11/42

"Many of the Germans are very sick, and the M.O. tells us that most of the German Army has dysentery. Some of them are so ill that Newton and I decide to move them up to the hospital. However, British O.C. troops move them back again. One poor man is very ill and cannot walk. He is thin and weak; the poor man, he must feel much worse now having been moved from the hospital back to the hot stuffy overcrowded pen. About midnight, a message is delivered to my cabin, 'Hoffman is dead'. War or no war I share the view that Hoffman was killed by the treatment, or lack of. I don't feel very well myself, but managed to get things going. We got a couple of crew to make a sack and prepare for the funeral tomorrow morning."

"The American Padre who can speak German, is also an episcopal like Hoffman, so he heads the funeral service and Hoffman slides over the side into the sea. All is over for him.

"Newton and I now insist on moving as many sick people as possible up to the hospital and, after this sad case we have our way. Also we open up another hold for them so they have more room. The more doubtful escapees have a hold for themselves where they can use a separate lavatory and eating quarters."

23/11/42

"Arrive at Colombo. There is evidence of some Japanese activity with a few ships sunk in the harbour. Colombo is warm and wet. We are not allowed ashore. I don't think we have missed a great deal. Our day, apart from seeing that the POWs don't escape, is to spend in sleep and playing shuffle board, a deck game! In the evenings we often spend the time talking to the German doctor. Doc. Feicer is treating Doc. Brown, the ship's M.O., for his upper lip that has been afflicted with some skin disease."

26/11/42

"Every morning Newton and I take turns to do a tour of duty. This consists of taking the M.O. down to visit his sick at about 9 o'clock and, while he attends the patients, we are supposed to see that no plans for escape could be made. Either plan or escape is more than I can understand, especially as there is so much sea between here and Germany. We usually look at the guards and then return to the upper decks. There is no cause for concern down there. All is under control."

Chapter Twenty Two
Guarding the Germans

WE HAD SET OFF FROM THE canals of Egypt on 16th November 1942, and first called at Colombo 23rd November, now we were on our way down to Perth and then Sydney, Australia.

Guarding all five hundred German P.O.W.s was actually fairly straight forward. We were on a ship, so I reasoned to myself that this reduced the chances of their escape. However, I realised that if there were to be any mutiny, this would in turn reduce

Photo c.1940 of the tightly cramped accommodation for the Allied troops, providing some idea of the crowded conditions on the lower decks, where the POWs were on this particular voyage.

our chances of escape. We developed a clear plan: the prisoners were to be kept in the lower decks, and so all that should have been needed was for a few armed men to stand at each entrance between the lower decks and the upper decks. Then we decided that these men would need some sunlight, so we formed a routine where every day the men were allowed up on to the top deck, to take in the sun, for an hour, albeit under guard. The German P.O.W.s really appreciated this, and the German doctor agreed, it would do a lot for their health.

It didn't take long before my men and I began chatting more and more to these POWs. We weren't supposed to do this, but coming from the RAF we were used to striking up conversation with strangers—usually first about planes, and then, eventually, about anything else. It was a very embedded part of the RAF culture to easily make friends with people and always be welcoming. In fact, it was often said that anyone from the RAF could walk into any air base and find a friend to talk to. This culture crept in when we were guarding these Germans. We certainly had never been trained to guard prisoners, and to us, these were men of the same age, with similar interests, who just happened to be on the wrong side of the war. They were also nearly all fluent in English, which was fortunate because most of us had not mastered the German language. Soon, quite a few of us from the RAF had struck up some good friendships with 'the enemy'. The British Army Officers were more disciplined in this area, and tended to not talk much to the Germans, but not so the Air Force blokes; my men were a bit of a rabble, however, you could absolutely trust them with your life. The RAF men weren't so interested in all the rules and traditions, they were from practical backgrounds and this came through in their approach to everything.

Amongst the Germans was a medical doctor whose name was Feicer, but being a POW, he had no equipment or supplies. However, he did not let this deter him. The doctor, staying in a cabin between the hospital, and mine, talked to me regularly, and he and I soon became good friends. Initially we had escorted him a bit more, but now, because he had earnt a bit of respect and freedom, he would often come up from the lower deck to play battleships with me and eventually we let him dine amongst the allies. We both enjoyed our time playing games, and it was a good way to pass the day.

The German doctor spoke perfect English and, despite his lack of medical supplies, word soon got around the ship about his resourcefulness and brilliant skills. The Allies heard about this and many of them started consulting with him rather than the ship's doctor. Apparently the British doctor, although good at medicine, lacked a decent bedside manner. Before long, most of the Allies, even the crew, preferred to see Doctor Feicer, who was a Nazi.

Before Hitler came to power, Feicer had a lady friend. The doctor explained that it was soon drawn to the attention of the affectionate couple that she had Jewish ancestry and so, sadly, he had to break it off with her. Feicer explained that he felt that there was no way of hiding her heritage, and no point in either of them becoming martyrs. After this, he thought the best thing he could do was to join the Army—perhaps it was to try and get out of Germany, or away from any possible repercussions, but in any case, he didn't seem particularly sympathetic to the German cause. He simply enjoyed making sick people well, and he didn't seem bothered about which side of the war they were on. He would have made Hippocrates proud and held firm to the idea that he would 'remain a member of society with special obligations to all human beings', and to him there was no distinction between races when it came to healing.

On route to Perth, many of the German prisoners were very ill with malaria, dysentery and all other manner of diseases. Many still suffered with bites all over their bodies from the desert flies that left an acid on their skin, which had now blistered and became infected. I watched the German doctor perform a few procedures on such sores, first on his fellow POWs, and then as he became more and more trusted, on the wounds of some of the Allied men. It looked as though he was using a razor blade to open the wounds and then bleed them and remove anything from the bites. He soon ran out of razor blades; he had almost nothing with which to treat people, except words of wisdom and the authorities were in no hurry to equip him.

As soon as we docked in Perth, on 3rd December, I took leave for a day or so and went ashore with the express purpose of collecting what I could from the chemist shops. I bought anything that I thought might be useful for a doctor and after visiting several chemists, I had quite a large collection, as well as a few boxes of razor blades. When I took them back, the German doctor was very grateful and apparently all the chemist supplies were very useful. I couldn't say exactly what they were used for and can only thank the chemist for loading me up with the proper things.

WWII Advertisement for Gem, a common brand of razor blades.

Before we had arrived in Perth, I had struck up a rapport with another German, a man who had been the navigator of a U-Boat. This friendship was based largely on my amazement with the man's skills, and I expect from his side, my approval revealed a lot to him. Furthermore, the friendship was also because I figured it was probably the best way to keep an eye on him; keep your friends close, keep your enemies closer. The U-Boat navigator had drawn a freehand map of the world, from memory. It was incredibly accurate and what amazed me more, was that everyday he would pay close attention to the sun and his watch. He was able to calculate exactly where we were, and each day he plotted our course on his hand drawn map. Then every night, the U-Boat navigator would seek me out, and show me his map. I would take it and show the navigator of the *Mauretania* and our navigator would then nod his head, almost regretfully, and certainly in disbelief. The German was dead accurate.

This meant that the five hundred German prisoners knew

exactly where we were, should they have wanted to take on the one hundred Allies who were controlling the ship. We all knew that this system of cooperation was anything but normal, but we thought it was much better to be honest and know how much each side knew. Amongst our guarded prisoners, navigation was not the only dangerous or useful skill—depending on whose side you were supposed to be on. The Germans had also calculated exactly how much oil was on the ship, and they knew how far she could travel with her current load.

After a short stay in Perth, we went south along the Great Australian Bight and on to Sydney. Here we picked up more POWs, this time the crew of a German Raider, an armed commerce raiding ship, that was supposed to be completely disguised as a merchant vessel, to allow it to pass undetected, but something obviously had gone wrong for these guys. It seemed interesting to me, that although they too were prisoners of war, the Raider crew were viewed more as civilians, particularly by the crew of the *Mauretania*, and so they always dined at the same tables as the Allied crew and then were free to roam the upper decks.

The *Mauretania* returned to Sydney Harbour, Australia, 19th August, 1945.

The captain of this German Raider ship was a good man and a strong leader. One evening he sat at the table I was at, and a group of us asked him about his experiences throughout the war. In perfect English, he explained that he had been on another ship off the coast of Western Australia and it was sunk; from there he was held prisoner, by the Japanese. I asked him what he thought about the Japanese, after his experiences, and he said, 'the Japanese have three sorts of enemies; the friendly enemies, neutral enemies, and enemy enemies'. He couldn't understand why we were busy fighting a war in Europe when he thought the world's efforts should be concentrated in the East. He didn't have a kind word to say about the Japanese.

The Raider captain and his crew had learnt of their fellow

German's navigational skills, and consequently we now had a ship with hundreds of prisoners, a full crew and accurate knowledge of exactly where in the Pacific Ocean we were. For some reason, despite the apparent risk, the Raider crew remained free and privileged aboard the ship.

Meanwhile, my twenty-five RAF men and I, together with just a handful of Army officers, concentrated on guarding the five hundred POWs on the lower decks. When I say guarding, I must admit, it certainly wasn't textbook style down there either. We had been issued with captured Italian rifles but they were unlike any rifles we had used. It became immediately obvious to the Germans that we had no idea how to use the things. I suppose being Italian weapons had something to do with it, but in any case, a few of the Germans seemed to know exactly how these rifles worked. Apparently one of my men stood at the entrance to the lower deck, armed, as he was supposed to be. Then a POW saw him fiddling with the rifle, and realised that this guard had no idea what he was doing. Rather than trying to take advantage of the situation, this prisoner taught and even physically showed the RAF man how to load and use the rifle! The German then resumed his position as a prisoner, walked down amongst the others on the lower deck, and continued being under guard.

In my spare time, while up on the deck one day, I asked the obvious question: I addressed the captain of the German Raider, and asked, 'Have you thought of taking over the ship?' He said, 'Of course! But we won't do it unless we can get close enough to a neutral port'. He continued, 'We know how much oil is on the ship and have calculated how far she can travel'. But this wasn't the only thing they had to keep in mind; the Germans did not want to go to any port that was Japanese occupied, and in the South Pacific, that ruled out a lot of options. The two nations, in our minds, may have been lumped in together as a part of the 'Axis', but as far as the Germans were concerned, returning to anything like their former experiences when they had been interned in Japan, was not an option. They knew that they would be much better off in an American POW camp.

Everyone seemed to relax even more, not that our guarding of the POWs had ever been particularly formal. As they wouldn't revolt against us, unless we were near a neutral port, the most important thing was for us to plot a course that was always far enough away from such ports, and ensure we didn't refuel with too much oil—as to not give them too many options for ports that they could potentially reach. I think in some places, they may have been able to make a break for it, and try to reach South America, but they knew this wouldn't have achieved too much, and there was no guarantee that they could have actually taken over the ship. We were armed—and now knew how to use the rifles! It would have been a blood bath. The Germans weren't silly, they weighed up their options, and decided that going peacefully to America was in their best interest.

From Sydney we went on to New Zealand, and spent a week there. The main purpose of this leg of the journey was to drop some New Zealand Army men home. Again, I went ashore and went shopping, this time mostly for Christmas supplies. I also arranged for the New Zealanders to provide some army escorts to look after the prisoners while my RAF men and I spent a few days leave ashore. The next thing I needed to do was to ask the British Army officers, who had continued conducting pay parades on the ship, if they could also pay my men for this fortnight. My men hadn't been paid since they left Egypt, and needed some money for their shore leave. The reply from the Army was a distinct 'No'. I am not sure how it came about that the British Army was able to pay its own men but there was no one onboard with the authority to pay the Air Force men.

As soon as I got into Wellington I went and found the New Zealand Air Force Office and asked someone if they could please pay my men. To my surprise they said, 'Yes', but I had to sign my name to it. There had been no one to ask so I just did what I thought was right. I had done this on my own initiative. I even told the New Zealanders that His Majesty's government would reimburse them at some point. It worked in the short term, but it did cause me some trouble later. We had a pay parade, where they called out each man individually, and everyone got paid. The RAF men were then able to go and enjoy themselves in New Zealand.

On the docks, I met a woman whose father was the

Cunard White Star Advertisement for the *Mauretania*.

manager of a local radio station. I was invited to go and have a meal at their house and she showed me around Wellington. The place was full of Americans, which the New Zealanders were not happy about. There seemed to be a lot of friction between the host country and the Yanks; mostly on account of the sheer number of Americans in Wellington at this time. While Japan was waging war all over the South Pacific, Wellington had become the first staging post and so the once quiet city was now full of Americans. I expect that the reason that we British were so well received, and were able to get paid so generously, was that the New Zealanders found it rather refreshing to have some Allies who were from somewhere else, for a change.

From Wellington, we went across the Pacific to Hawaii. With careful navigation and planning, we never got so close to a neutral port as to tempt the Germans to raise a mutiny. We did not need to travel in convoy, because the *Mauretania* was large enough, and fast enough, to not be threatened by any lurking U-boats, but this did however, raise the stakes if the Germans attempted a revolt, we would be completely on our own.

This leg of the journey covered Christmas, so the POWs decided to put on a show. It was wonderful, especially considering that they had almost no materials to work with. Nonetheless, they managed to find a collection of odd things which they turned into Christmas decorations, and together with the decorations I had purchased in Wellington, we soon had some well decorated lower decks! I am not sure that I was supposed to provide our prisoners with Christmas decorations, but it seemed harmless. Some of the prisoners were still very sick, but the men all pulled together; I think they felt that they needed this to lift their spirits and the show was more for their morale than anything else. In fact, it was only the guards who ever saw it, all the other allies remained on the upper deck having their own high teas, fancy dinners and dances. The Germans had worked together as a team, and managed to produce the most marvelous Christmas performances, we even exchanged 'bon-bons' and lollies.

Being Christmas, there was also a fancy dress party for all those who were on the upper decks—the Allies, our crew, and the crew of the German Raider. Apart from those in the forces, there were also a few other very important people on board, including the Victorian Governor and his wife, although I think they remained oblivious to the celebrations that were going on with the POWs. It was quite an upper class affair. After the party on the lower deck or, more formally I should say, once I had finished my shift guarding the prisoners, I found my way towards the bar on the upper deck. I was chatting to one of the men, which I tend to do whenever there is dancing going on, hoping not to be noticed by anyone.

A short while later, a woman approached me and asked if I would care to dance. I said the first thing that came to my mind, 'I'm terribly sorry but I can't, it's because of my wooden leg'. I felt she was far too competent a dancer for me. She said, 'That's fine, I'll sit and join you for a drink instead'. I realised that I was getting into deeper trouble. She was not going to let up, and I think she felt that she was being rather charitable to sit and keep me company. Then I found out who she was; she was the wife of the Governor of Victoria. So from that moment on, wherever I was, whenever I saw her, I had to start limping. It was very embarrassing for me, and no good for my back, which actually was still rather painful, and which would have been a more obvious excuse, had I thought before opening my mouth. Some of the men thought it hilarious when they found out what was going on. Lies will find you out, there is no doubt about it.

There was also a very senior Air Force man onboard with two small children who were about five and eight years, and they seemed rather fond of me, so whenever I wasn't busy with the prisoners or organising my men, I looked after them and played games. We got on very well, and I enjoyed their company, so it was suggested, that every time there was an event on the ship, I would look after them. This worked out well for their parents, and brilliantly for me, as it prevented any more potential dances or stories with the Governor's wife.

While we were in the middle of the Pacific Ocean, it was decided that the army—both those serving as guards, and those on board in all manner of other capacities, would like to have a parade. I am really not sure what the purpose of this was. My men were more 'technical' and 'practical' and in truth they were thought to be a rather shabby bunch. I was told to inform the RAF that everyone was doing a parade, but that not much was expected from them and that they were to be put at the back anyway. When I informed my men, rather than responding as though it was 'just another parade', they were indignant; they all decided to make it the best parade they had ever done. Knowing that the Army blokes were full of self-confidence and that they expected to show themselves up compared to the RAF, my men put in their greatest effort. We discussed the best way to do this, and how we could rehearse without tipping off the Army. I was to lead their section of the parade, something that I had never done before. They told me to just 'shout a bit of something or other' and we will do the rest. We continued planning and organising our best parade ever.

We hadn't completely practiced and I was still a little unsure about how well we could pull it off, but the day finally came. There were all these army characters up in front, with an audience of some of the big wigs that were on board, including the Governor of Victoria and his wife. The RAF men filed in and stood at the back. There they were, with their best dress on. They had all their buttons polished perfectly and their shoes were shining. The RAF men had spared no effort in doing their best parade ever. Everything was perfect. The parade was held on the top deck of a ship in the middle of the ocean. It was a pity there weren't more people to see it, the RAF would have been proud. We made the army look silly; compared with us, they looked like an unruly rabble! We showed them up in every way. I don't think the senior ranking men in the Army ever forgave us! We found it amusing, and worth all the effort.

Soon we were scheduled to arrive in Hawaii, which I learnt first from the German U-Boat navigator! On our way in to Pearl Harbor, we could see the *Arizona*—now only exposed partly, and although work had begun to remove her frame, she was still not totally sunk either. Surprisingly, the Americans wouldn't let us off the ship. The Germans thought this was hilarious; my men thought it was ridiculous. I'm not sure why they wouldn't let us off, but I suppose they were very nervous about us opening the main entrance with a few hundred prisoners onboard. On the docks, alongside the ship, many local girls danced for us. They wore beautiful looking grass skirts, and hardly anything to cover their tops. There were hoots of merriment from all the blokes on board. In exchange, the men would throw coins down for the ladies.

A few Americans came on board and I spoke to a couple of them. They were very interesting, and rather nice, and I enjoyed talking to them.

Not surprisingly, as we stood there chatting, in Pearl Harbor less than a month after the first year anniversary of the attacks, they too, didn't have a kind word to say about the Japanese.

From Hawaii, we sailed to San Francisco. By this stage, we

One of many American propaganda posters designed to encourage U.S. citizens to become more involved with the war and encourage more volunteers.

were very comfortable with the POWs. Many friendships had been formed, which in any other context, would not have had the time or the means to form. We had enjoyed an uneventful journey, and over the few months, the men had built up a mutual respect; we just saw it as befriending men with similar interests, who were of the same age, and who just happened to be caught on the wrong side of the politics. The Americans, however, did not share this view.

When we arrived in San Francisco and docked, as soon as the gangplanks were put in place, the Americans marched on board, armed to the teeth, and blowing their whistles. They were so militant and stern. They assembled the Germans and marched them off in groups of ten at a time with two armed guards for each group—one at the front and one at the back. While the prisoners were waiting to be marched off, officiously in groups of ten, they had been shaking hands and wishing us well. Many onboard were also exchanging addresses and making promises to catch up after the war had ended. So, as the Germans left the ship under armed guard, in small groups, they would stop, one by one to shake hands with the British and say, 'We'll see you after the war'. The Germans and most of us on our ship found this to be most amusing. Only a few moments before we had been fraternising and exchanging farewells, and there was nothing the Americans could do about it. I can't imagine what they expected of the Germans; there wasn't anywhere for the Germans to go, and being an Allied continent, I couldn't see that running away would have been a likely option.

Chapter Twenty Three
Homeward Bound

AFTER DOCKING IN SAN FRANCISCO and bidding farewell to all the prisoners, I needed to look after my men, and get them back to England. I had not been briefed on what to do when we arrived—in the land of Uncle Sam—but it was clear that this was the end of our journey on the *Mauretania*. I went and found someone from the American Air Force and asked if they would be able to pay us, and once more I promised in the name of His Majesty, that someone would reimburse them. I gave them my word and I even signed my name to it. I couldn't see any other way of getting all of us from San Francisco to New York if we hadn't been paid. The Americans did agree to pay us, which was good of them, but they weren't terribly generous. With a few US dollars, we managed to buy train tickets for the journey from West to East, the idea being that once we arrived in New York, we would be able to board a ship for England.

Chicago railway station. A scene typical of any US railway station during WWII as servicemen bade farewell to loved ones.

The train journey was pretty relaxing. We travelled in a large group, all the RAF men and the few Army blokes who

Originally published in 1916, this image of Uncle Sam went on to become, according to its creator, James Montgomery Flagg, one of the most recognised posters in the world
It was extensively used for recruitment and propaganda throughout WWII.

had been guards with us on the *Mauretania*. I am not sure where the rest of the British Army dispersed to once we arrived in San Francisco, but I suppose, since most of them hadn't been on the ship for the purpose of being guards, they must have been sent to America for other duties. Travelling with the group, all of whom had become comfortable and familiar friends, we had an enjoyable journey; with the exception of five or six men who had become quite unwell. These men suffered from the awful fevers and chills, which we all recognised as malaria relapses.

I made acquaintances with the train conductor; he was a good sport and told many stories about the train, and various passengers he had met over the years. We stopped periodically, but didn't usually get off the train. One of the more lengthy stops I remember most clearly was at Salt Lake City. No one from the military got off the train, but because we were going to be there for a long time, the train conductor decided to jump off and go and visit a friend who lived in the town. After an hour or so, the train was ready to go again, but the conductor had not yet returned. I looked out of the window as we started pulling away from the station, and saw him frantically running along the platform. He was trying to catch up to the train, or perhaps he thought that by flapping about and making as much of a scene as he did, that he would get the attention of the driver. The train certainly didn't slow down. Fortunately, as he ran along the platform, he was able to grab hold of a handle and pull himself aboard; it was like a scene from an action movie. Once he had settled and caught his breath, and walked up the train to our cabin, he explained that he had made this stop many times before to visit this friend, but this time, while he was away, the train had moved to another platform, something which had not been factored into his tight timing.

As soon as we arrived in New York I went looking for an American Army base, in the hope that they could provide some accommodation. I would then need to visit the British Consulate in New York to arrange tickets for a ship to take us back to England. The men from the Army base were very good, and gave me directions. Step by step, in the loud and slow voices they reserve for foreigners or those hard of hearing, they told me go so many stops on the subway and then which subway exit I should take; I was told that if I did this, I would come up on Wall Street. I followed my instructions, caught a train and at the appropriate station I walked up onto the street level. It was just an ordinary little narrow city street. I asked the first bloke I came across, 'Excuse me, can you please tell me where Wall Street is?' The man looked at me and said, 'Say bud, you're standin' on it'. It was not at all like I had imagined.

I spoke to someone from the British Consul and told them the story: who we were, where we had come from and how many passage tickets we needed to get across to England. He looked at my identity card, and puzzled when he read the statement on the back; either impressed or confused with what it said, he just as quickly looked at me and said, 'I'll take your word for it'. He was very helpful, but said it would take a few days to organise. I also managed to have the men paid again. It wasn't very much, but they were grateful for the few dollars.

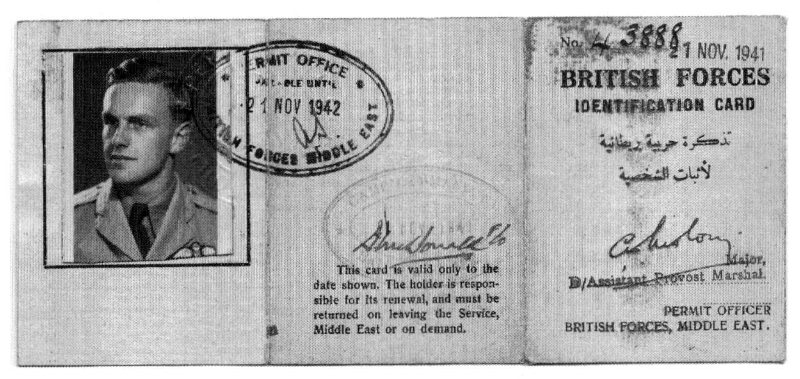
PERMIT OFFICE
VALIDE UNTIL
21 NOV 1942
BRITISH FORCES MIDDLE EAST

This card is valid only to the date shown. The holder is responsible for its renewal, and must be returned on leaving the Service, Middle East or on demand.

No. 43888 21 NOV. 1941
BRITISH FORCES
IDENTIFICATION CARD
تذكرة حربية بريطانية
لأثبات الشخصية
Major,
D/Assistant Provost Marshal.
PERMIT OFFICER
BRITISH FORCES, MIDDLE EAST.

The front of Cyril Johnson's Identification Card, on the back of which (see below) one third of the card was filled with personal descriptors such as height and eye colour, and a statement declaring his immunity against Egyptian Criminal and Civil Courts.

IMMUNITY — EGYPTIAN CRIMINAL AND CIVIL COURTS.

1. Under the provisions of the Convention dated 26th August 1936 to Article 9, Treaty of Alliance, Egypt, no member of the British Forces in Egypt shall be subject to the criminal jurisdiction of the Courts of Egypt, nor to the civil jurisdiction of those Courts in any matter arising out of his official duties.

2. The holder of this pass is a member of the British Forces, Middle East.

3. This Immunity from the jurisdiction of the Egyptian Courts does not, however, imply that offences may be committed with impunity against the Egyptian Laws or against the person or property of residents in the country, but merely that these offences will be dealt with by the Appropriate British Courts which have been constituted for the purpose.

4. Should the holder of this pass receive a summons to appear before any Egyptian Court for any criminal offence he or she will formally inform the Court that immunity from the jurisdiction of the Court is claimed in accordance with Article 9 of the Treaty of Alliance.

A criminal offence is one for which a penal award by way of imprisonment or fine can be made and includes contraventions of any sort.

5. Immediately the holder of this pass receives a summons to appear before any Egyptian Court he or she will report to Headquarters, B.T.E. or Headquarters, R.A.F., Middle East, through the usual channels full particulars of the charge and the court to which summoned together with confirmation that action as in paragraph 4 above is being taken.

6. If the holder of this pass is used for damages or debt, in a 2 Court of civil jurisdiction in Egypt, and such claim is concerned with a matter arising out of his or her Official duties he or she will at once report the case through the usual channels to Headquarters, B.T.E. or Headquarters, R.AF., .Middle East. (C.R./Egypt/80983/A).

A few days passed, and we were told to go down to the docks and board the ship that matched our tickets, a very small Belgium vessel, with a peculiar Belgium name. The ship, if you could call it that, had been built in peacetime for passenger cruises along the Meuse, a smooth river. Consequently, it had a very flat and shallow bottom. It was certainly not built for a winter voyage across the North Atlantic Ocean! Neither was it designed to take half the load it now tried to keep afloat. We boarded and walked down to our deck, stunned at how many men were already squashed onto the boat.

Being an officer, I was one of the very few men to be given a cabin. I looked around to check where my blokes were sleeping and some of them were actually quite sick at this point. One man, in particular, was very ill and obviously distressed about sleeping in the common rooms alongside everybody else. We decided that it wouldn't do to have the sick men sleeping down on the lower decks where it was so crowded and poorly ventilated; the German POWs had better conditions on the lower decks of the *Mauretania*. We got some mattresses and put them on the floor in my cabin. I didn't know how else to care for them and nobody else seemed interested. There wasn't a lot I could do, but at least I could offer them somewhere safe and warm for the night.

The word safe played on my mind. I wasn't so sure that anywhere on this boat was safe. As we made our way across the North Atlantic, the boat rolled about furiously. It was unlike anything I had ever experienced. Her flat bottom provided no stability and we progressed very slowly, rocking greatly with each wave. Things were bad on the boat, and our only reassurance, if you could call it that, was the fact that we were travelling in convoy. This made me feel that if our ship should actually capsize, the others would rescue us, but I soon realised that the reason we were travelling in convoy, was because we were facing another, more significant risk. Across the Atlantic lurked many U-Boats. These posed the most significant risk; their weaponry could easily sink us. We were literally 'sitting ducks' to a U-Boat. Unlike the *Mauretania*, this Belgium vessel could not out run a U-Boat—in these waves she was struggling to even steer her own course—but if she were torpedoed, I expect she would take on enough water to sink well before we could get most people up from the over crowded lower decks. It was quite a scary thought, so I tried not to ponder it too much.

Fortunately, pondering anything soon became out of the question; I had to focus on the immediate. A storm had picked up. The boat had been in danger of rolling before the storm and now the movement on the boat was phenomenal. It wasn't long until the sway of the boat became so violent that even the casings over the brass steam pipes in the companionways broke loose. People were literally in danger of being knocked over by them as they rolled along the corridors. Then the smaller upright piano in the main lounge started moving and as the swell continued to grow, the grand piano started to move. It quickly gained momentum and went about like a tank, careering all over the room, crashing from one wall to the other and back again. People were starting to get frightened; I was terrified.

One of the tankers travelling in our convoy was quite close to us at this point. One minute it was there, the next minute it was gone. It had just suddenly broken up, taken on a lot of water and disappeared. It took seconds to sink. Sadly, we weren't able to rescue anyone from the tanker and I think it would've been much too dangerous to have circled around to try again; I am not even sure that we had any control over where we were going at that point. Everybody onboard was doing whatever they thought best to protect his or her own life. We could no longer take comfort in any illusion of safety offered by travelling in convoy.

There was now a great line up on the lifeboat deck, where many people had started gathering. We were absolutely terrified. I did a few mental sums and realised that there were not enough lifeboats onboard to cater for everyone. There were about five or six hundred of us squashed onto this little riverboat, so we would have been a few hundred short. In any event, I wasn't sure that people would even be able to stay in a lifeboat in such wild waves. The convoy had been scattered. It was every boat for itself and every man for themselves. Just walking on the boat and trying to make our way to the upper deck was difficult and we had to hold on to anything we could.

When I made it out onto the top deck, rain was gushing down in torrents, and the fierce ocean spray made it even wetter and colder. The wind lashed about violently and the noise was deafening; people were shouting instructions at one another, but with the rain it was difficult to see what they were mouthing. One second we would be sky high, and the next we would crash down between the waves, and see large foreboding walls of water taller than us. Each one looked as though it would swamp us, but just at the critical time we would be raised up, on another peak, and then begin our rough descent again.

We stood on the deck, with people gathering around the lifeboats, and a few people started to get pushy. Fortunately, almost as suddenly as it had picked up, the storm started to calm. People just stood there, not really wanting to leave their place in the queues for the boats. We had survived unharmed, and it actually took a few minutes to realise that it really was calming down. Eventually we made our way back inside, where it was clear that we were now in desperate need of a port for some serious repairs, so we made a stop in Reykjavik, Iceland. Most of us onboard were from North America, or the UK, so we thought we knew what cold was, but this was something else; January on a wet and windy ship in Iceland.

To lighten the moment, and offer some comic relief, an American Colonel on board decided to share the story about his pal, an American pilot. Apparently, this pilot had been flying somewhere in Canada and made a forced landing. For whatever reason, he had ditched into the ocean and needed to swim a long distance back to shore. The problem was that he had a cork leg and he soon found himself in strife with one floating leg. The American Colonel was a dramatic storyteller, and helped us conjure up all sorts of images of this poor man with legs flailing and his head below the waterline, struggling to counteract the natural tendencies. Not usually particularly funny, but in this context, when we were so tense, his humor and acting skills were greatly appreciated and drew a large crowd of tentative but relieved passengers.

We eventually arrived in Liverpool. It was now late January, 1943. Many of us on board had not been home to England for a couple of years. We felt so excited to be coming into port. When we docked, we were surprised that the customs officers came aboard the ship. They systematically started to look through everyone's belongings. Then, they declared that we would need to pay taxes on anything that we were importing. We couldn't believe it! Here they were, with their safe jobs, not going off to war, and they had the arrogance to act as though we were in peace times.

The customs officers made no exceptions. For most of the RAF men returning from Africa, buying and wearing gold jewelry had been the only way they had of saving their money over the last few years, but this wasn't even considered. The men had collected little items from around the world, particularly gold watches, necklaces and coins as a way of keeping their wages safe, and on them at all times, because we didn't always know if or when we would be returning to our bases and banks across the central African desert were in short supply!

I don't think any of my men had enough cash to pay the import taxes they were being quoted. In any case, it wasn't the lack of money that was the main problem, it was that the men were so outraged. It was ridiculous that the customs officers behaved the way they did. In some cases, their jewelry was the result of two or three years wage, serving the crown in foreign nations. Many of the men found ingenious ways of hiding their belongings, but as for the larger items, rather than give the customs officials the satisfaction of confiscating them, they simply threw them off the boat and into the sea.

I got off in Liverpool, without much to my name, and saw to it that the men were able to buy their train tickets and make their individual journeys home. I am not sure what I was supposed to tell them to do, but I gave them leave; it had been a long time and it seemed like the proper thing to do. I too went home to see my mother, whom I had not seen since July 1941.

I headed home, unannounced and unexpected by my family. When I arrived, it was about one o'clock in the morning, and of course, everyone was in bed. I was surprised to find that for the first time in my life, my mother had locked the front door of our house. As far as I could remember, it had always been unlocked. I wasn't even aware that we had a key for the door. As I couldn't just let myself in, I threw some little pebbles up to her bedroom window, on the second floor. She awoke and came rushing down.

The first thing mother did was to apologise that the door had been locked; she then caught herself and realised the weight of moment and hugged me, with the longest embrace of my life. Later, she informed me that she had only started locking the front

Aida Monica Cooke, Cyril's mother, sitting by the fire in the piano room.

door a few months earlier, when one night she had accidentally gone to sleep in the afternoon with the reading lamp on and the curtains open. She awoke startled to find a policeman inside the house. He had seen her reading light from the street and because

Lilly Moss walking with Cyril Johnson, on leave from war, 1943.

the town was supposed to be in a total blackout, he had simply let himself in to draw the curtains. Apparently, he was in the process of switching off the light for her when she awoke, and suggested that, perhaps, it was a little too easy for someone to break in and that she should consider locking the front door, especially on the nights that she was sleeping in the house alone.

The next day, I decided to go into town. I was chuffed to note how much was the same as it had always been, and I stood there, on the street corner, under the Aspull Fingerpost, and suddenly everything felt right and familiar. For a minute I could almost believe the war was over, but there was still plenty to remind me that such was not quite the case. I made my way into the town, and noted the quietness of the place. I walked past a solicitor's office, and a little way down the street I stopped to read some of the messages on the town notice board and the British propaganda posters. The next thing I knew an excited young woman came running out calling my name. She had been working as a legal secretary, and as soon as she saw me, her boss said, 'Right'o then, pack up and have the afternoon off'. It was Lilly!

We enjoyed our time walking around the town and sharing a meal together. I was so happy to be home, and this just made it incredibly special, I had my head ten feet up in the clouds. We spent a lot of time catching up on all the news, and sharing what we knew about our various childhood friends who had been posted abroad, but she spoke more and more about one particular friend, Harry, a young lad who had lived in our village, and who I had attended boarding school with. By the end of the week, I knew my luck had not changed and somehow the 'subject' of Harry, was raised during nearly every conversation—Harry this and Harry that... He was a good man, and I couldn't begrudge him anything.

I enjoyed a few weeks leave, but I soon got a demand from the Air Force to come down to London and explain my actions. They had received a bill from New Zealand for the wages of my men. They had also received another one from San Francisco and again from New York. I explained that, although I had been travelling with the British Army, they had refused to pay my men, and the men needed something to live on, over the four months between Egypt and Liverpool. The Air Ministry wanted to know how I had managed to get these men paid. They reminded me that I had no authority to have done something like this, and that I hadn't followed the rules. I said, 'Yes, but there was no one else there so I did my best, and anyway I didn't know what the rules were'. I got out of that okay, but the wages bills for my men had been quite a substantial amount of money and they certainly wanted me to account for it, or at least squirm a little before they let me off.

Chapter Twenty Four
Back to School

IN MARCH 1943, I WAS sent to Watchfield Beam Approach School where I learnt to fly aircraft purely by instruments. The day I arrived, I didn't think I'd be expected to do any flying because there was thick fog everywhere. Nonetheless, I presented myself at the school, but they said, 'You're just in time, you are about to go out in just a few moments'. Then I was taken by car across the field to our aircraft. It was very difficult to find the aircraft because the fog was so thick, we could hardly see anything at all. We eventually found the plane, and I thought, 'He's not going to start this blooming thing up', but he did.

We taxied out in the thick fog, with a lead car to guide us. I couldn't see a thing, but we went and flew a circuit. The instructor flew the first circuit while I just held on for dear life. Part way around the circuit, my instructor turned to me and said, 'Now this is what we are here for, to teach you instrument flying, in a big way'. The idea was that we would just go out with the instructor and learn how to fly using instruments to determine our height and position, and not need to rely on any visual cues, at all.

I suppose, in a way, I had previously come to understand the basics of this at least in theory, but until now, I had always done my best to avoid flying in cloud. Indeed, on the few occasions that I had needed to fly through cloud in Africa, things had gone horribly wrong, and on more than one occasion, planes from my convoys had crashed and exploded into the hills. I was, therefore, rather cautious about flying in cloud or fog! It's difficult to imagine, until you first experience it, but one of the most dangerous things about flying without seeing the horizon

Airspeed Oxford, formation flying with a thick cloud in the background.

is how easily one can start to roll, or even completely turn upside down, and still be certain that they are flying straight and level. The Beam Approach School was to teach me to overcome this apprehension and trust these newer and more reliable instruments, and fly without being able to see the horizon or any visual aids. More specifically, it was to teach me to fly in total darkness, with *deadly* accuracy.

The idea of Beam Approach was to find a signal that a pilot could latch onto; the pilot would receive audio feedback, enabling him to tell from the signals exactly what direction he was headed in, relative to the transponder beacon. There were short notes, 'dots', if the aircraft was pointing left of the beacon, and longer sounding 'dashes', if it was pointing right, and when he was on the correct track, the dots and dashes merged to a constant signal. There was a whole series of different beacons, and once a pilot latched onto one, they would adjust their heading to make sure they were flying towards it. Then, as the pilot flew over the beacon, a different note would sound, and they could latch onto the next beacon, confirming an accurate heading, and at the same time make any necessary adjustments to their height.

When we were coming in for a landing, our vision was often totally impeded by thick fog. It was impossible to tell that there was even an airfield below, but the beacons were there, and once a pilot found one, it was easy to adjust to the correct height and find the inner beacons and after that, the specific runway markers. Finally, the poor pilot would be flying at a very low height, only just above the ground, and at the last second, he could see the runway only a few feet below. Sometimes even after we landed, the fog would be so thick that we still couldn't see the runway. The system was incredibly accurate; the pilots were not always so accurate. The issue was learning to trust the instruments and fly blind. It was all about trusting your instruments and learning to ignore any natural instincts. I trained mostly on twin engine Oxfords, which I had used in my early days at flying school—where we had the privilege of seeing the horizon! We started off by doing figure eights in the air to get used to lining up with the transponder beacons. Later, as we got more confident or competent, or perhaps just sillier, we were encouraged to fly further away from base, maybe a few miles, and then turn around and try to find the airfield, somewhere amongst the whiteness.

Aerial view of Watchfield Airfield, on a rare clear day.

When we weren't in an aircraft with an instructor, we had a 'link trainer' almost the same as we had used in the elementary flying school. It was similar to an aircraft cockpit, but mounted on a large mechanical pedestal. The cockpit could be made to

NO. 1 BEAM APPROACH SCHOOL
R.A.F. WATCHFIELD.

COURSE No. 183
DATE COMMENCED 5/2/43 COMPLETED 9/2/43

TOTAL HOURS ON COURSE

FLYING	DUAL	SOLO I.F.	1st or 2nd PILOT	PASSGR.
DAY	12.10			
NIGHT	—			

TOTAL FLYING 12.10
LINK TRAINER 5.00

ASSESSMENT

CHIEF FLYING INSTRUCTOR.

DATE 9/2/43

No. 1 Beam Approach School, RAF Watchfield, Certificate acknowledging 12:10hrs instrument flying time, and 5hrs in the Link Trainer.

move around, and for those inside, it felt as though we were actually flying. It was so realistic, that even with the years of experience we had, one of my classmates became air sick. It was an excellent way for us to practice, because we had to go through all the same processes as instrument flying, and the instructors could also record what we had done and how accurate we were, without the dangers of actually flying blind.

I did slightly over twelve hours training on instrument flying, before someone was willing to paste the following certificate into my log book on the 23rd March 1943:

"This is to certify that Cyril has received instruction on the following practices, and has carried out the said practices unaided: circuits and landings, flapless landings, single engine flying on both port and starboard engine to include feathering, single engine

landings on port and starboard landings, over shoots, steep turns and evasive action."

I was given two additional certificates to paste into my log book, which, "*Certified that I have been instructed in and understand the Wellington IC fuel, oil and hydraulic system, and operational endurance,*" and after this was signed, I was given a second memento which, "*Certified that I and my crew have been instructed in and practiced Dingy Drill and Abandoning the Aircraft Drill.*" Looking back, it was a rather short time in which to learn so much, but the war felt more urgent than ever and the RAF needed us to be trained and able to fly night time operational sorties as soon as possible.

From there I did another short course: Bomber Command. The first thing I needed to learn was to fly a much larger aircraft; a Vickers Wellington.

Barnes Wallis, who must have been a genius, designed this. It was geodesic in its construction—with the members of the fuselage being curved rather than in straight lines—which meant it would hold together more easily in the event that some pieces should be missing! Perhaps an insight into what was to come for some of the pilots.

Being taught how to fly the Wellington was similar in many ways to how we were taught to fly every other new aircraft. I spent a short amount of time with an instructor, and from there, it was over to me to fine-tune my ability to control it, and get some experience under my belt. I was shown how the instruments and controls worked, and how to handle the aircraft in different positions, and all the stress limits. Mind you, they didn't emphasise the stress limits; I found it absolutely amazing to see what we were expected to do with the machines.

After this, I learnt how to use the aircraft as a weapon. It was a specific course in which pilots learnt how to do all the maneuvers they would need to do and the roles their crew would play in order to complete a successful mission. This part of our training was run very well.

We were taken to a larger base where we would meet people from all the other Bomber Command training schools. There were new navigators, bombers, radio operators and

A pre-war formation of No. 9 Squadron Wellington Mk I aircraft.

gunners. The people were all from different schools and weren't 'sorted out' by the administration staff. The various training schools had produced the right number of different people, but the officials thought that rather than pre-allocate the crews in set groups, they would let us work it out for ourselves. We all gathered in the mess, and were told that if we wanted to, and if we met people that we got on well with, that we could give our names, and together we would become a crew. This worked rather well, and people tended to sort themselves out. Given the pressure of the situations we were soon to face, it was important that we select personalities that we would work effectively with; teamwork would be imperative.

Wellington Bombers lined up at the RAF Stradishall, ready to fly to Brussells and Paris as a demonstration of strength. Note the scale of the aircraft compared with the airmen.

Our superiors must have been feeling the urgency of the hour, as my log book shows that on the very day we chose our crews we went out and did some cross country flights, specific target practice and carried out simulated attacks over a specified empty field. The following evening, on 21st April, 1943 I had an instructor, for one flight only, because it was my first ever night flight. I flew with my instructor's crew, and then he left me to my own devices, with my new crew.

Construction of the Geodesic Airframe of the Wellington, comprised of over 1650 parts.

Damage to a Wellington, resulting from a direct hit from anti-aircraft fire, near Duisburg, Germany, April 1943. Despite the loss of the rear turret and its gunner, the pilot was able to fly the aircraft home to the UK, highlighting the resilience of the aircraft, and the remainder of its crew.

Following this, I was sent in a Wellington to Seighford for my next course, the '21 Course'. I was supposed to do training there but two days later I was sent on to Hixson instead, where we undertook more training, focusing on night flying, with only a bomb aimer and radio operator, but no navigator, probably because this was just local training, and because I expect there was a larger volume of classroom work for the navigators.

We spent a lot of time in a portable Nissan Hut, being lectured and briefed on all that they thought we would need to know for night flying and operations over enemy territory, and there was a lot of vitally important information to take in. Listening to the lecture on how and where we were expected to fly, was enough to send shivers through my core, and it was probably made all the more surreal by the nonchalant manner in which the information was presented. If ever there had been a time to pay attention in class, it was now. It was expected that we would hear it once, write it down and know it; we couldn't rely on the important things being repeated or drilled into us. The volume of critical information was too great, and there was not enough time to have the luxury of such spoon-feeding.

To complete my training, it was a prerequisite to go on at least one real mission, a proper sortie over the enemy. For some reason, mine was cancelled at the last minute, and they forgot to reschedule me again for this part of my training, so I would just have to work it out on my own, which I was prepared to do, and they signed me off.

Interestingly, by the time I had officially finished the '21 Course', the RAF had begun training pilots on the newer aircraft, the Halifax, which was a four-engine aircraft. The Wellingtons were very good operational aircraft, and had been the dominant bombing aircraft during the first years of the war, but the Halifax was much larger, with seven crew, and it was much more powerful with the capacity to carry a greater load of bombs. Rather than the initial plan for me to be operational on Wellingtons, in November 1943 I was sent to a conversion unit where I learnt to fly the Halifax. It was strange timing, because it seemed as though all my training on the Wellingtons went only towards learning to fly the Halifax. Although, I must admit, it helped to have learnt to fly the Wellington first, and in the months that would follow, I would be grateful for all the experience and training I had accumulated.

The Halifax was a four-engine bomber, similar to the Avro Lancaster, but made by the Handley Page firm. The earlier series of Halifax had radial engines, which at night meant the ring around where the exhaust came out would glow slightly. For targets in difficult conditions, the Halifax planes were less appropriate, because they were more visible, with four faint but glowing circular bulls eyes against the night sky. Apart from that they were a beautiful and comfortable aircraft, albeit they were a lot larger than what I was used to.

Group photo, providing a scaled perspective of the size of a Whitley bomber. (Ten Squadron at Lemming Bar).

The first time I flew the Halifax solo was quite an experience. As I was taking off down the runway, there was a flock of birds sitting at the end. I didn't appreciate that they were there until I had built up speed, but I made the decision to take off anyway, because I definitely did not have enough runway in front of me to be able to stop. By then, seeing a roaring metal bird with a 104 feet wingspan coming towards them, most of the birds had made the decision to take off too! I thought it would be ok, as for a second it looked as though most of them had left, but suddenly, the last few birds who had failed to take off with their flock decided that now was a good time to clear off. I collected several of them and one got stuck in the air inlet of an engine; I had already left the ground, so there wasn't much I could do. It is never a great start to be forced to shut down an engine on take off, but I had no choice, I completed a circuit and landed as soon as I could.

During our training time we were also sent on some genuine operational exercises. One flight we went on a search mission, because someone had ditched an aircraft over the sea, and it was our task to find any survivors. We couldn't find anything. Come to think of it, I am not sure that I really achieved much for His Majesty with my time on the Halifax. For me it seemed to serve more as a transition for me to learn to fly the next aircraft. Many others went on to fly Halifaxes operationally, including a childhood friend from Wigan, but in my case, it seemed that as soon as I was competent on an aircraft, rather than sending me out to battle, they signed me up to learn to fly something bigger or faster, or both.

I think it may have had a lot to do with the timing of when I was going through these courses, but the next thing I knew, I was being instructed in the Avro Lancaster. The two aircraft were similar in many ways. It was a very brief conversion course. The Lancaster was again a four engine aircraft. It had inline Merlin engines, which didn't glow at night, so they remained less visible. According to my logbook, I did only two flights under instruction. Both were at night, and mostly just circuits for familiarization. Then I was sent to lead my crew, on my first real operational trip in any of the aircraft over the enemy or at night; I was headed straight for Berlin.

Chapter Twenty Five
Bomber Command

OVER THE COURSE OF WORLD War II, Bomber Command controlled the RAF bomber forces. At the beginning of the war, it was agreed that aerial bombing, from both the Axis and the Allies, would be restricted to military targets and ports, railroads and some infrastructure. However, after the Germans bombarded Rotterdam, Holland, in May 1940, the RAF was authorised to retaliate. Bomber Command was instructed to bomb German cities of particular military significance, regardless of the inevitable civilian losses. The gloves were off. The idea was to force the Germans into a position of defence rather than attack. Specifically it was a strategy to shift the focus away from the production of offensive weapons and toward defensive ones, and to wear down the morale of the German people.

It is often said that Bomber Command, and its accomplishments, formed some of the most strategic and heroic aspects of the entire Second World War, with19 Victoria Cross medals being awarded to members of bomber crews. However, two generations after the war had ended, critics of history have stood back and judged these actions. I neither seek to defend nor condemn the strategic decisions that were made from on high, but can only say that at the time, things looked very different to how they may present in the light of day, on the pages of a history book. There are many resources thoroughly detailing and critiquing Bomber Command, and two that have particularly shaped my understanding of the events that unfolded are, Hasting's 1979, *RAF Bomber Command*, and Saward's 1984 *'Bomber' Harris, the Authorised Biography*, but here I will provide a brief account, as I understand it, for the purpose of contextualising my experiences. I do not claim to be a historian, but I think the facts and figures speak for themselves with regards to the magnitude of the situation and the enormity and complexity of the decisions faced by all involved.

A decade after the war ended, the tactics of Bomber Command would be summed up in a statement by Hitler's Minister of Armaments, Albert Speer, "*The real importance of the air war consisted in the fact that it opened a second front… Defense against air attacks required the production of thousands of anti-aircraft guns, the stockpiling of tremendous quantities of ammunition all over the country, and holding in readiness hundreds of thousands of soldiers, who in addition had to stay in position by their guns, often totally inactive, for months at a time.*" Speer went on to say, "*No one has yet seen that this was the greatest lost battle on the German side.*" However, it certainly came at great cost to the allies.

Members of 576 Squadron in front of a Lancaster Bomber, providing a scale for her wingspan. Photograph taken in 1945.

Perhaps a picture of Bomber Command is best created through the appreciation of some statistics, as the scale and magnitude of these operations is otherwise difficult to comprehend. The individual aircraft themselves, the Avro Lancaster, arguably the 'building block' of Bomber Command, was enormous. Many squadrons had their group photos taken in front of a Lancaster, and her wings would easily cover a row of 70 people standing shoulder to shoulder.

The Lancaster wingspan was 102 feet, and she was just shy of 70 feet long, capable of cruising at 275mph with a full load, at an altitude of 24, 000 feet for 1,730 miles. By 'full load', I mean seven crew, 2,150 imperial gallons of fuel and up to 18,000lb of high explosives, or if the turrets were removed, a single 22,000lb 'Grand Slam' bomb. This particular bomb was referred to as the

'Ten Ton Tess', and was 26 feet, 6 inches long and almost 4 foot wide.

Despite the Lancaster's grand size and power, the operational life of this heavy bomber was an average of only 40 hours flying time. Many of the losses came from the German night fighter planes, and to counter this, two .303 machine guns were fitted in the front turret, another two in the upper mid turret and a further four in the rear turret. A single Lancaster was capable of doing a lot of damage, but we could not afford to be over confident, because the Germans certainly fought back hard—the inevitable consequence of the Allied strategy, to force the Germans into concentrating their weapon production and war efforts on the defensive.

There was nothing small about a single aircraft in Bomber Command; never mind how enormous it was as an organisation. Bomber Command involved so many people and so much complex organisation that it was structured into Groups 1 to 6, under each of which were smaller sub-groups: base, station, squadron and then flight. Each squadron may have had around 20 operational aircraft, and there were hundreds of squadrons. It is difficult to say how many aircraft each squadron had, and the exact types, due to the constant losses and replacements of aircraft after each night raid. A single operational airfield may have a strength of around 2500 personnel, who were all directly or indirectly involved in crewing or supporting the crew, servicing, loading, maintaining and preparing the air craft and bombs, as well as photographic recording teams and navigation, communication and briefing teams. In North Lincolnshire alone, there were eight such operational airfields, each with a similar number of personnel. The scale of Bomber Command was enormous.

During the latter part of the war, night after night, the phrase 'maximum effort' would accompany the operational orders from Head Quarters. It became an almost nightly occurrence to have several hundred aircraft flying over German territory, dropping bombs and getting their share of the enemy's counter attacks. This resulted in an incredible overall achievement: 364,514 allied sorties dropping almost one million tons of high explosives on the enemy. No one doubts that this was a critical factor in the outcome of the war, but again the cost to the Allies was immense.

Out of 125,000 Bomber Command crew 55, 573 were killed; an incredible 44.4% death rate. Additionally a further 8,403 were wounded, and 9,838 became prisoners of war. Over the course of the war, on each night mission, bomber command sustained an average of a 2.2% death rate. However, between November 1943 and March 1944, nightly sorties flown over Germany sustained an average death rate of 5.1%.

To compound these risks, it was determined that to complete a 'tour,' crews must complete 30 operational sorties. This meant that on the official statistics, less than 27 out of every 100 men would survive a full tour.

Before I knew all this, I was sent from my training school to North Lincolnshire, England. The airfield I would operate out of was Base 13, Elsham Wolds. At this stage of the war there were two RAF Bomber Command Squadrons operating out of this airfield, 103 Squadron and the one I was to join, 576 Squadron. Squadron 103 had flown a total of 486 operations, consisting of 5,480 sorties or 'flights', many of which were part of the Battle of the Rhur and the Battle of Berlin.

The Royal Air Force Ministry of Defence would later publish an article entitled *576 Squadron*, on their website, from

which I was reminded that the 576 Squadron was much younger, and was formed as a part of Group 1 at Elsham Wolds, with operations beginning on 3rd December 1943, only four weeks before I arrived, and ultimately disbanded in April 1945. We all flew Lancaster Mk I and Mk III aircraft. Our squadron's motto was Carpe Diem—'Seize the Day', and our badge was distinct from other squadrons, depicting a merlin, with its wings inverted, preying on a serpent. The 576 Squadron flew a total of 2,788 sorties, and mostly flew operations as a part of the Battle of Berlin up until May 1944, but unfortunately, this was over the period in which Bomber Command were sustaining the shocking average of 5.1% loss on each mission.

Chapter Twenty Six

Operations – Berlin

IN MY LOGBOOK ARE MANY pre-set columns for the date, aircraft, name of pilot and passengers, duty and remarks. Typically each flight is recorded as a single line entry, and in red ink for night flying. On 1st January 1944, I made a normal, one line entry, part way down the page, which says in the duties and remarks section: '*Operations—Berlin*'. That was it. There was no fanfare about it. Looking back, I can now say that there was rather a lot more to it. At the time this wasn't something that I really had words for, and it certainly wasn't something I

No. 576 Squadron . Elsham Wolds.

Year 1944 Month	Date	Aircraft Type	No.	Pilot, or 1st Pilot	2nd Pilot, Pupil or Passenger	Duty (Including Results and Remarks)
–	–	–	–	–	–	— Totals Brought Forward
JAN	1	LANCASTER	[illegible]	W/C Rollins	Rollins Crew & Self	OPERATIONS – BERLIN
JAN	3	LANCASTER	P2	Self	Crew [illegible]	X Country
JAN	5	LANCASTER	V2	Self	Crew	OPERATIONS – STETTIN.
JAN	7	LANCASTER	P2	Self	Crew	Bullseye as ordered.
JAN	8	LANCASTER	P2	Self, Sgt [illegible]	Crew [illegible]	F/A & Bombing
JAN	14	LANCASTER	F2	Self	Crew	OPERATIONS – [illegible] Abortive. [illegible]
JAN	28	LANCASTER	B2	Self. Sgt Reed.	Crew, [illegible]	Bombing, F/A, Air to Sea firing
JAN	30	LANCASTER	B2	Self	Crew [illegible]	OPERATIONS — BERLIN

Summary for JANUARY 1944. Unit 576 Sqn 'B' Flt. Date 31/1/44. Signature Cyril A B Johnson F/Lt. Aircraft Types: 1. LANCASTER 2. 3. 4.

S/Ldr. O.C. 'B' Flt.

W/C O.C. 576 Sqn.

GRAND TOTAL [Cols. (1) to (10)] 603 Hrs. 30 Mins.

Totals Carried Forward

Excerpt from Cyril Johnson's logbook, record and summary for flights in January 1944.

wanted to write much about in my diary, but I remember it with haunting clarity.

This was to be my first ever sortie, given that the one I was supposed to do as a part of my training on the Wellingtons had been cancelled and never rescheduled. Being my first probably didn't mean too much to the other airmen because all of them had only been there a short time themselves, and there was such a high turnover rate of crews. Crews lost in action were constantly being replaced, and I realised that I was just one of a never-ending stream of newcomers. However, the fact that this was my first sortie certainly meant a lot to me; there was so much to observe and learn.

It was New Year's Day, but around the airfield the day shift were all busy working and preparing for the evening's operation. The operation they were getting ready for wasn't any different to any of the other 'maximum effort' attacks they had been used to preparing for, night after night. There was a lot of activity going on all around the place, amongst the ground staff, while most of the night crew were sleeping—many having just returned from an all-night sortie in the early hours of the morning, and many knowing they would be sent out in a few hours time.

As each Lancaster sat there with her doors open, over the course of the day, more and more people would call past and perform their duties, ensuring the planes would be ready for takeoff late that afternoon. First, the armament teams would come and replace or repair any guns that needed work after the previous night's efforts. Soon people from the navigation and communication teams would visit and ensure the equipment was all serviced and ready for another flight. The photographic teams would also inspect each plane, checking that the proper cameras and films were ready and set to take pictures at pre-determined intervals after the release of the bombs, an important way of determining the levels of success the missions were having in total. By this stage of the war, things were now in well-established routines, and much of the groundwork was fast paced and reasonably efficient.

Meanwhile, in the offices, many of the staff from the Women's Auxiliary Air Force, or 'WAAFs' as we called them, and liaison officers from other services, had spent the morning preparing a report on the results of the previous night's bombing effort. They checked photos from the bomber aircraft's automatic cameras, plotting on a map the release points of each of the returning planes. After examining these dark images with a magnifying glass, their findings would be compared to the additional photos taken by the reconnaissance team who flew over the target at first light. The photos would be compiled with the results of the interviews done when crews had returned that morning and the information would then be compared to previous efforts and maps. Photos of targets yet to be destroyed would also be brought out, and all the information would be sent to the briefing room where the top brass seniors would analyse the data.

The completed report would then be sent to Bomber Command Group No. 1 Headquarters, and an appropriate course and set of targets for the evening would be determined. Based on the distance to the determined targets the Chief Operations Officer would prepare an order and issue exact instructions, detailing the fuel quantity and the specific bomb loads to be carried by each aircraft. These secret instructions were then sent by the teleprinter, an electromechanical typewriter, over secure lines, to all the relevant airfields. All of that had been taking place behind the scenes, and away from where I stood, looking around Elsham Wolds, thinking about what the night may have in store. Sleep didn't come easily in those daylight hours that led up to my first night time sortie.

Later that day petrol was delivered, with a large tanker backing up and parking under the wings of the planes. The Lancaster's wings contained the fuel tanks, which held just over 2,000 imperial gallons. We had heard that, from the evidence gathered after investigating crashed planes, the Germans knew precisely where to aim their cannons, but it was of little comfort to dwell on that, or shift my gaze down towards the bomb hull and the thousands of pounds of explosives that I would be riding into the night sky.

The drivers of the tankers had been told the exact quantity of fuel to deliver to each aircraft. On this day, it was to be a full

Image taken at Scampton, Lincolnshire, depicting the manpower required to keep a Lancaster operational. First row (left to right); flying control officer, WAAF parachute packer, meteorological officer, seven aircrew; pilot and captain, navigator and observer, air bomber, flight engineer, wireless operator/air gunner and two air gunners. Second row: maintenance crew; N.C.O. fitter, flight maintenance mechanic, N.C.O. fitter, five flight maintenance mechanics, electrical mechanic, instrument repairer, and two radio mechanics. Third row: bomb team; WAAF tractor driver with a bomb train and, following that, three bombing-up crew: fourth row, ground servicing crew; corporal mechanic, four aircraft mechanics, engineer officer, fitter/armourer, three armourers, radio mechanic, two instrument repairers, three bomb handlers, machine gunbelt fitter. Back row: AEC Matador petrol tender and two crew, Avro Lancaster B Mark I heavy bomber, mobile workshop and three crew.

Caption and photo courtesy of
http://ww2today.com listed as 3rd December 1943, 'orchestrated hell'.

load. In theory the only ones who knew the target for the night raid were in headquarters, but the experienced petrol operators could calculate the flight time from knowing the quantity of fuel together with the weight of the bombs. Most times, they could surmise the target, just from these facts. That wasn't such an issue, but the neighbouring farmers had also discovered this. It was often said that if you wanted to know where you would be sent that night, that you could just ask one of the farmers, as they could give you a good idea of at least the number of hours flying time you had, and make comments like 'Eight hours... probably off to the capital tonight'.

While the fuel was being pumped, over in a distant corner of the field, the bombs were being made ready for loading. Fortunately this part, known as 'the dump', was well away from aircraft and all other buildings and concrete blast walls also protected it. A tractor, towing a train of three or four, open, flat trailers with guide rails at various distances, arrived ready to be loaded.

Avro Lancaster with two trailers loaded with bombs, forming the tail of the bomb train. Note the man sitting on the bombs.
(This aircraft is "S for Sugar" which had an Australian connection during its 137 sorties and is today housed in the RAF Museum, Hendon, UK.)

With the use of large specifically rigged pullies and chains, a group of armament men began gently swinging large 500 pound general bombs towards the first trailer. At this stage the bombs were considered safe because they had not been fused, so the men were not worried about walking one along each side of the bomb to guide it down on to the trailer tracks. They loaded two bombs at a time, with a total of six bombs per trailer. These 500 pounders were torpedo shaped with fins to stablise them in the air and each 500-pound bomb carried 250 pound of highly explosive amatol powder. These were likely to be what our plane

Bomb load with a 4000lb 'cookie' and 12 bomb containers, each loaded with incendiaries, in this instance, 236x4lb incendiary.

Bomb bay of a Lancaster. Note the man reaching up and ensuring the 4000lb 'cookie' is correctly situated.

would carry that night, as they formed the majority of most planeloads; there were however, a few other combinations of odd sized bombs that the Lancasters frequently carried.

The armament team then hoisted a 4000lb amatol bomb up with a tractor to be gently placed on a trailer of its own. This was known as the 'cookie' or 'dustbin' due to its shape; a blast bomb, designed to explode on impact, and consequently it had such a thin shell, that 75% of the weight of the bomb was the explosive powder. After the cookie had been gently placed, this trailer was hooked onto the two other loads of 500 pounders. Eventually all the bombs on a train of several trailers would be taken to the hut and fused, many with impact fuses, and some with time delay fuses. They were then all towed, very carefully, to just one Lancaster, who would usually carry 14-18,000lb of explosives, depending on her fuel load.

WAAF driving a Fordson Tractor towing an unusually long bomb train.

Across the airfield I could see many tractors operating at a time, all pulling a train of three or four trailers carrying various bombs to load a single aircraft. When they arrived at a plane they were manhandled, with a high degree of delicacy and precision, and winched up into the aircraft. This was a dangerous job for the armament department. Although it was mercifully rare, there had apparently been occasions when, due to mechanical and electrical faults, some bombs had dropped, with lethal consequences, and so it was with the greatest respect and care that each bomb was loaded.

Each Lancaster carried one of several set combinations of explosives. For those given the code name 'arson,' they carried fourteen small bomb containers, each loaded with 236 four-pound incendiary bombs. These small bombs were designed to burn and burn, with little that could extinguish them. The load represented by the code 'plumdoff' consisted of one 4000lb 'cookie', three 1000lb short delay bombs, and up to six small bomb containers each filled with either thousands of four pound, or sometimes, hundreds of thirty pound incendiary bombs. Other aircraft carried fourteen 1000lb bombs—each with a mixture of fuses, some set to explode on impact, and others time delayed to up to six days later, all designed to cause maximum harm to the enemy.

The Lancaster was also capable of carrying a single 22, 000lb (ten ton) 'grand slam' bomb, depending on the distance she would need to fly, because the fuel load would need to be reduced and adjusted according to the additional weight. A large fuel load, with a light bomb load, meant long range and on this New Year's day, each Lancaster was being loaded with around 14,000lb of bombs. At this stage it was still not officially known, but the word around the field was that there was fuel enough for a long flight, so the target would be Berlin. Despite all the secrecy, I had no reason to doubt the airmen. They seemed to know what was going on, and they had a few more weeks practice at interpreting the clues.

It was by now late afternoon and I made my way back to the briefing hut, where a few hundred men soon showed up, some on the backs of trucks, others on their bicycles. They all came from various parts of the airfield, which meant riding long distances as the field extended for several hundred acres. Our bicycles were of great importance, to us and to daily life on the airfield. My bomb aimer, a Canadian named Candy Gurus, was a clever young chap, who I watched riding in and noticed that he was ambidextrous enough to roll a cigarette and cycle at the same time.

My crew was fantastic, and I felt reassured to be flying with them. They all knew their jobs well and they had all faced many high-pressured situations before. A few months earlier, while we were still in flying school, when the authorities had given us the opportunity to choose our own crew, we had all put our names forward, and now I was thankful. When we had first met all the people from the various Bomber Command schools, I had two Canadians approach me, one was the bomb aimer and the other was a rear gunner. They asked if they could join as a part of my crew. We then had two English men, a radio operator and navigator, and an Irish mid-upper gunner approach us. We all got on rather well and formed a solid team. We actually got on so well that when we were not operational, we would often all go out for a drink together.

I caught up with them quickly before the briefing. Everyone seemed alert and refreshed, perhaps they had slept better than I had, or perhaps it was just adrenalin. Gurus stood there puffing away on the cigarette he had just rolled, while the other Canadian, a man named Webster, who we called Webby, started chatting about how he hoped everything would be in good order with his guns for the night. He was in fact the rear gunner, and one of the most passionate people I had ever met when it came to guns. From the moment I had first met him a few months earlier, it was immediately obvious that he had come from a backward part of Canada and had grown up all his life with guns. For the moment, the other boys were pretty quiet, particularly the Irish mid-upper who I don't think said a word.

Typical briefing hall for Bomber Command crews. This image is of the 408 Squadron, RCAF.

It was now only a few hours before the flight, and we had been sent to the briefing room. We entered the room as a rowdy young bunch, all talking to one another. At the end of the room was a stage, upon which there was a huge map, marked with the details of the exact route we were to fly, but when we first entered, the map was covered with a black curtain. The base, station and squadron commanders walked in, and suddenly all the men would stand to attention, until these officers had filed past and taken their seats. The main briefing officer was very dramatic. He would start his speech, and then he would swiftly remove the curtain and uncover the map. On this occasion there was a great groan from everyone. This had confirmed our suspicions, and Berlin was not a popular trip. We more or less knew what we were in for. The briefing officer didn't try to rev us up. That

Poster prepared by the British as a part of the Careless Talk Costs Lives campaign, in which a German night fighter plane is tipped off to be on the offensive.

would have gone down like the proverbial brick, and besides, we weren't American. Rather, the whole briefing process was taken seriously, people were usually pretty quiet and solemn, but we certainly didn't need any hype.

Everything was supposed to be a secret right until the last minute. The RAF didn't want any word getting out about where we were heading on a mission, as they didn't want the enemy to have any time to concentrate their defenses along that route. There were no rumours going around our airfield to suggest that spies were about, but I suppose it was always possible. Apart from that, the authorities were worried that someone could have said something out of turn, or perhaps that word may have spread around the town. They were understandably always very edgy about people saying anything to anyone. The commanders tried very hard to stop information from getting out. Just prior to a briefing, they even shut the gates to the airfield so that no one could go in or out, but, for all of this, there was nothing to stop the local farmers looking over the fence.

A major part of the briefing was being shown the map. On it was the course that we were to fly—a jagged route—because if we had flown in straight lines it would have made it too easy for the enemy to predict where we were heading. The RAF would tell us exactly where we were going, what opposition we were likely to encounter from the ground, and in the air, and all the information that they thought we needed to know. The weatherman was always far too optimistic, but we got used to that. They told us what they thought the defenses in that area were like, but this wasn't always accurate. Actually, come to think of it, knowing how strong the defenses were, this wasn't too useful anyway—there wasn't much we could do about it, we just had to take it as it came.

Everyone was briefed together and each member of each crew took specific note of the things they needed to know. All the navigators were madly jotting down their detailed instructions and copying the route from the wall map precisely on to their charts. The radio operators learnt at that stage what codes or things they needed for communication. As pilots we had to take note of pretty much all of the briefing, but our crew were very good, and helped tremendously with recording all the information.

A part of the briefing was when to start and stop windowing; this was the code for a tactic designed to disrupt German radar. It worked when aircraft dropped thin strips of foil from the aircraft. They would be various lengths, but usually around 2x12 inches. When it was first used on 24th July 1943, it proved to be successful, with the drifting clouds of metallic strips scrambling German radar, and concealing the locations of the aircraft. Unfortunately, by now, the Germans had worked out how to filter out the spurious radar echoes, so it had little effect.

Every allied aircraft also carried a 'very pistol', which contained coloured flares that we could fire off if we got lost and needed to identify ourselves as Allies. This may happen as we tried to return home, particularly if we were much earlier or later than the rest of the squadron, and it may be the only way to identify to our own side that we were friendly and should not be shot at! There were specific 'colours of the day' that changed with each night, such that an ambitious enemy aircraft could not copy them. As a part of our briefing we were informed of the colours and told to check that we were carrying the correct colours for that night. I could think of nothing worse than making it all the way home and being shot down by my fellow countrymen, because we were carrying yesterday's flares!

The Flying Control Officer would then brief everyone on the takeoff procedure. The challenge was to get all the Lancasters, from both the 576 squadron and the 103 squadron, started and taking off in as little time as possible, without wasting fuel or spacing the aircraft too far apart. This was no mean feat in and of itself, but all of this had to be coordinated without the use of radio so the enemy couldn't listen in. Those who were later in the queue would have to take off in the dark.

The Station Commander would also emphasise the various cities en route and their defenses. After this he would move on to the importance of timing over Berlin. To achieve a concentration of about 600 aircraft over a city, within 20 minutes, all in the dark, without using radio, was a challenge! However, the concentration, if achieved, was designed to overwhelm their defenses, and thus provide us with increased odds, and I think this was supposed to give us a sense of comfort. The timing and altitudes prescribed were precise; most of the focus of the briefing was on timing and synchronization and to that end, the briefing officer would then count down, three, two, one, now, and a few hundred of us would synchronise our watches precisely. We walked out knowing that many whom we had just shared the briefing with would not return, but it wasn't something that we dared talk about; it just wasn't the done thing.

It was then time to make final preparations. Getting dressed was a bigger event than one would have imagined, but we were flying in poorly insulated aircraft at high altitude. Indeed, temperatures of –40° Celsius were commonplace and on really cold nights they could be as low as –60° C and of course, being in an aircraft, we couldn't jump about to get warm. I don't think it was too bad for us pilots, because we at least had the benefit of some warm air flowing in from the engine, but for the rest of the crew, conditions just got colder and colder the further back they sat.

This was especially true for the poor rear gunner, who only had a layer of thin Perspex between him and the elements. After one flight, our rear gunner got out and accidentally dropped

Bomber Command crew dressed in their full flying kit
(Crew Details unknown).

his uneaten orange on the ground; the orange shattered into a million little ice crystals.

We had thick woolen jumpers and then most of us had a large jacket with a canvas outer, which we wore as a middle layer. On top of this jacket we would wear our Irvin sheepskin jackets. If we wanted to, we had the option of wearing an electrically heated suit, which we could plug into the aircraft, exactly like an electric blanket inside a suit. Not surprisingly, with the number of risks we were already taking, wearing this suit and thus being wired into an aircraft, which itself was a 'flying bomb', did not appeal to many. We also had fur-lined boots, which most people wore with silk socks and then woolen socks on top to prevent frostbite. A few of us discovered that you could cut the boots near the ankle line and leave yourself with a pair of shoes, and then open up the leggings, and wrap the fur-lined material around your waist to provide some extra warmth and in my case, padding for my back. Despite all the layers, we still felt freezing cold as we sat still for hours at a time.

RAF issue fur-lined flying boots.

We went to get our flight kits, which were maintained and

Bomber Command Flying Suits.

stored by a specialist team of WAAFs, and many of the specific items in these needed to be signed out. Our silk parachutes were the most important. Most of the crew wore ones of the chest type that could in theory be strapped on at the last second, only when needed. This was to enable them to crawl about the aircraft, and perform their tasks in the already tight spaces they had to work with. Pilots however, wore full harness parachutes, for the entire flight. These looked like a backpack and formed a part of our seat. Most of the pilots complained that they felt like sitting on cold stones, but in my case, the unnatural curve of the pack made my back ache like blazes, especially by the end of a flight.

Crews being transported to their awaiting Lancasters.

We were also issued Mae West life jackets, in the not too unlikely event that we might have to ditch over the sea. Further, we were issued with whistles that were attached to our sheepskin jacket collars, so that we could find other members of our crew, should we need to ditch in the North Sea—and not freeze to death—as well as a variety of other useful and well thought-out supplies in our kit bag.

Once we were ready, the airmen assembled casually on the lawns by the main buildings for a few minutes and a last smoke. We huddled together in groups of seven, according to our crew, some clutching their lucky charms, others jumping up and down on the spot to keep warm or dispel their nervous energy. It was usually about this time we would tease each other the most, and the banter became the strongest. It was incredible how close everyone became with the members of their own crew, and how boyish we were in our displays of affectionate mockery and mild torment while we waited nervously. Our crews were our families; I suppose we knew that the odds were that we would all live, or all die together. Indeed, a crew would usually all fly together for every sortie, although the chances of them completing a tour of thirty, were less than one in three. This certainly shaped the way we bonded and got on with one another. By this stage it would be about an hour before the first of the Lancasters would take off, and there was still much to do. Soon we were transported from the lawns of the briefing hall to our plane, in the back of various little troopies and trucks because the aircraft were dispersed over such a great distance.

Chapter Twenty Seven
Bombing the German Capital

IT WAS NOW LATE AFTERNOON on 1st January 1944, just two weeks before my 24th birthday, and I was about to go on my first ever sortie. We were dropped off at our Lancaster. The aircraft had been sitting on the ground, and the engines had been tested earlier that afternoon by the ground crew, but already it was cold. When we climbed inside, I was not thrilled to see that everything was frozen, with actual ice crystals on much of it and condensation running along the parts we managed to warm. The aircraft was almost painfully cold to sit in, and initially we could not see through the windows.

I made my way along the low and cluttered narrow

The Captain running up the motors just before takeoff, in Lancaster G for George, as he prepares for takeoff on the flight from Prestwick Aerodrome, to Australia, on her 91st 'Sortie', c.1944

Inside G for George, looking forward from behind the main wingspan, on the left sits the wireless operator and on the right is where the navigator works.

Inside a Lancaster B Mark III, the flight engineer checks his settings from his seat in the cockpit.

Mid-upper gunner in his position, as seen from navigator's dome near the front.

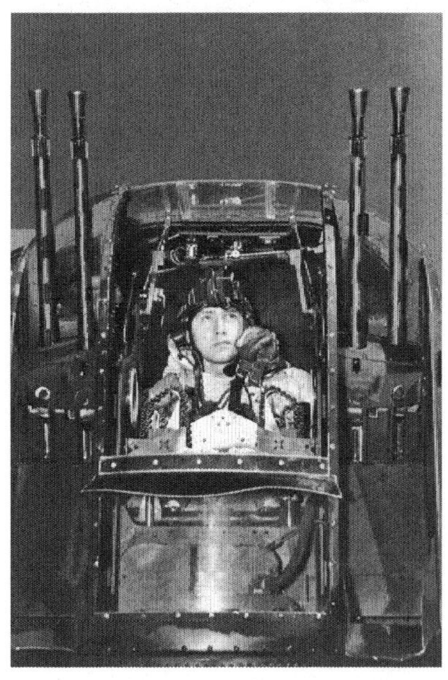

Rear Gunner in his position, with only a thin sheet of perspex between him and the outside world. c.1944, 630 Squadron.

passageway to the cockpit near where the flight engineer sat. Behind me was my navigator, an English chap named Woodfine, and nearby would be the radio operator. Further out sat the mid-upper and rear gunner and bomb aimer. Where I sat was perhaps the warmest place of the aircraft, having the benefit of air from the warm engines blowing in. The only place on the Lancaster, which was reinforced, was the area around the pilot's head, so that was a mercy too. Unfortunately, the down side to where the pilot was located was that I had to first climb over a rather large and obtrusive mains power box to get there. Had I not been in trouble with my back, I would say that this was only a small challenge for the average young and athletic pilot. However, I did wonder, how, in the event that we were struck, and presumably spinning out of control, a pilot was ever supposed to climb over this obstacle in the dark, avoid the tangle of oxygen hoses and find the escape hatches. Nonetheless, it didn't do to dwell on that idea either. There were a lot of things that one couldn't afford to think about in Bomber Command and at this point, I needed to focus on just getting into the seat with my back giving me as much grief as it was.

The ground crews were highly efficient. They would remove the necessary covers and assist with the start up process. They used a large external battery, wheeled in on a trolley, to reduce the demands on the internal batteries as the four large Rolls Royce Merlin engines were started. One by one, they all started, and the noise was incredible. I was thankful for my helmet and internal earmuffs. The cotton wool, which had not done my hearing any favours in Africa, certainly would not have been an option here.

I released the brakes and we rolled out to join the rest of the Lancasters from the 576 Squadron and the 103 Squadron. We were not able to use any radio communication to coordinate the takeoff but we were well aware of which signals to watch for and where to join the appropriate queues. We waited along the perimeter tracks for quite some time. I never knew how the pecking order was determined, but some Lancasters were fortunate enough to be in the front of the take off queue, which meant they took off in daylight. On the short winter days, the rest, like myself, started rolling after last light, something that made takeoff even more hazardous.

There were time pressures on us from every direction. All Lancasters needed to be coordinated in such a way as to make the takeoff process smooth, not too close together but not far enough apart that the others who circled around mid-air waiting, would waste their fuel. Petrol was precious and conserving it may make all the difference for a lost or damaged aircraft trying to limp home. Also, a close gap in the succession of takeoff was needed to enable as many aircraft as possible to take off while it was still daylight. If we took off too closely together, there would be no room for error, and no chance of a 'go-round' in the event of an engine failure.

Even without the time pressures or the challenges of coordinating so many aircraft without the use of a radio, the actual process of take off itself was hazardous. The perimeter track was about 50 feet wide, which meant that we all travelled in single file. Steering, though not usually too difficult, had to be done with precision. If an aircraft had wandered off the tarmac, her wheels under the weight of a full load would undoubtedly become bogged in the marshy grass and collapse the undercarriage. Apart from the potential hazard this posed with 14,000 pounds of impact fused explosives under my seat, at the very least, an undercarriage collapse would block others and cause great delays in take off.

Eventually I taxied out on to the runway; the aircraft felt as heavy as my burden of responsibility. I ran through my checklists again, everything needed to be done perfectly. The red light held solid while the aircraft just in front of me cleared the runway. Green. I opened up all four engines. Alongside the runway stood a group of ground crew and some WAAF to wave us all good luck, and I hoped it was not goodbye. We appreciated their effort, given how cold the evenings were. I turned my complete concentration to keeping the machine on the runway at high speeds. The flight engineer would call out the speeds, so that I could focus on maneuvering the aircraft. We broke through 100 miles per hour, well before the end of the 2,000 yard runway and at around 115 miles per hour I pulled back on the nose. All 65,000lb became airborne; we were now officially a flying bomb! Thank goodness we didn't have an engine failure at that point, as there would have been little option, and I dare say an absolutely tragic result. The weight of the aircraft pushed all stress limits, but soon we were up, and on our way to Berlin.

The first planes at the head of these missions were the Pathfinders. They had taken off well before us, and would go and identify the target, and drop brightly coloured flares over it. From then on, the theory was that all the other hundreds of aircraft coming from various airfields around the UK would converge into a stream and as they got closer to the target, they would eventually be able to see its boundaries. Unfortunately, it wasn't always so simple. The pathfinders were very skilled, and managed to find the towns most times, which was remarkable given that the cities were in total blackout. Meanwhile, however,

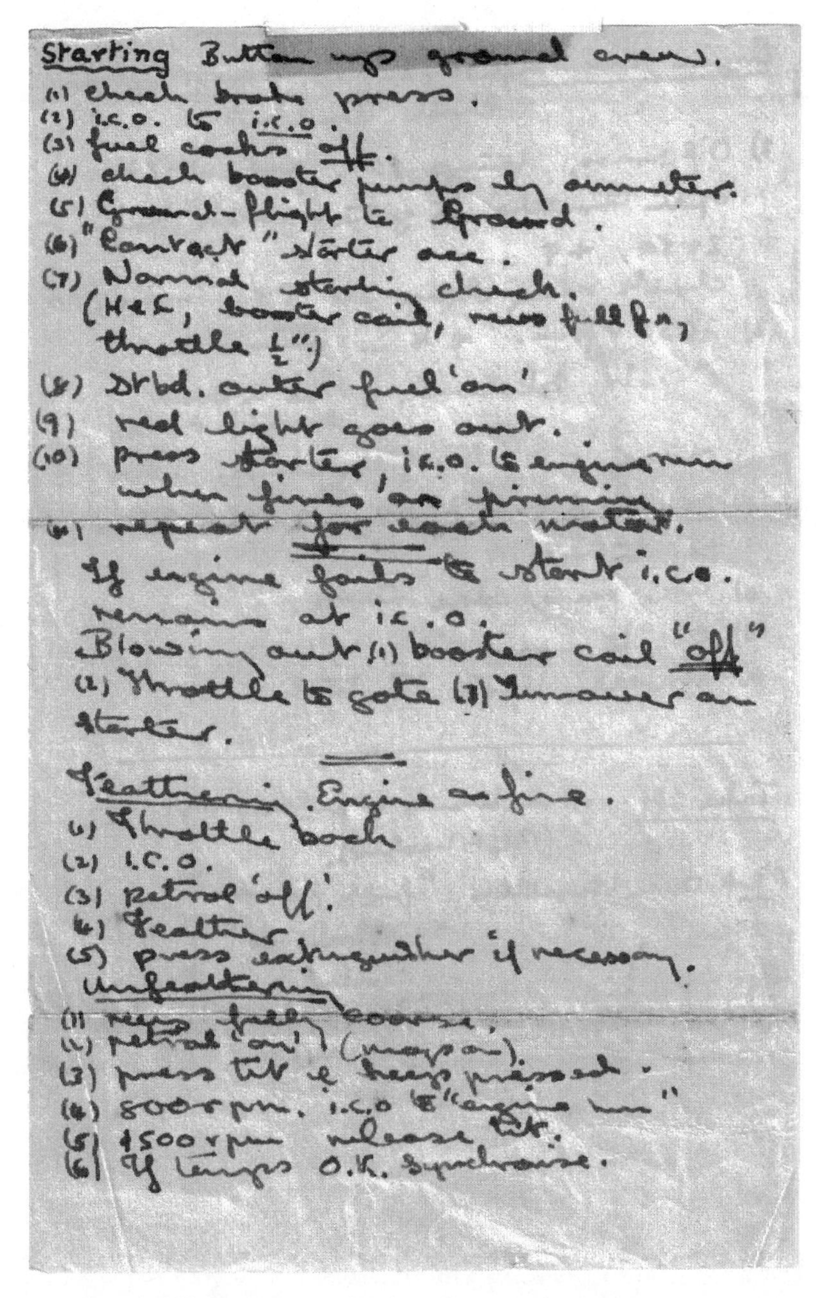

Starting Button up ground crew.
(1) check brake press.
(2) i.c.o. to i.c.o.
(3) fuel cocks off.
(4) check booster pumps by ammeter.
(5) Ground-flight to ground.
(6) "Contact" starter acc.
(7) Normal starting check.
(H&L, booster coil, revs full fn,
throttle ½".)
(8) Strbd. outer fuel 'on'.
(9) red light goes out.
(10) press starter, i.c.o. to engine run
when fires 'on priming'
(11) repeat for each motor.
If engine fails to start i.c.o.
remains at i.c.o.
Blowing out (1) booster coil "off"
(2) throttle to gate (3) turnover on
starter.

Feathering. Engine on fire.
(1) Throttle back
(2) I.C.O.
(3) Petrol 'off'.
(4) Feather
(5) press extinguisher if necessary.
Unfeathering
(1) revs fully coarse.
(2) petrol 'on' (mags on).
(3) press tit & keep pressed.
(4) 800 rpm. i.c.o to "engine run"
(5) 1500 rpm release tit.
(6) If temps O.K. synchronise.

The takeoff checklists Cyril wrote out and committed to memory.

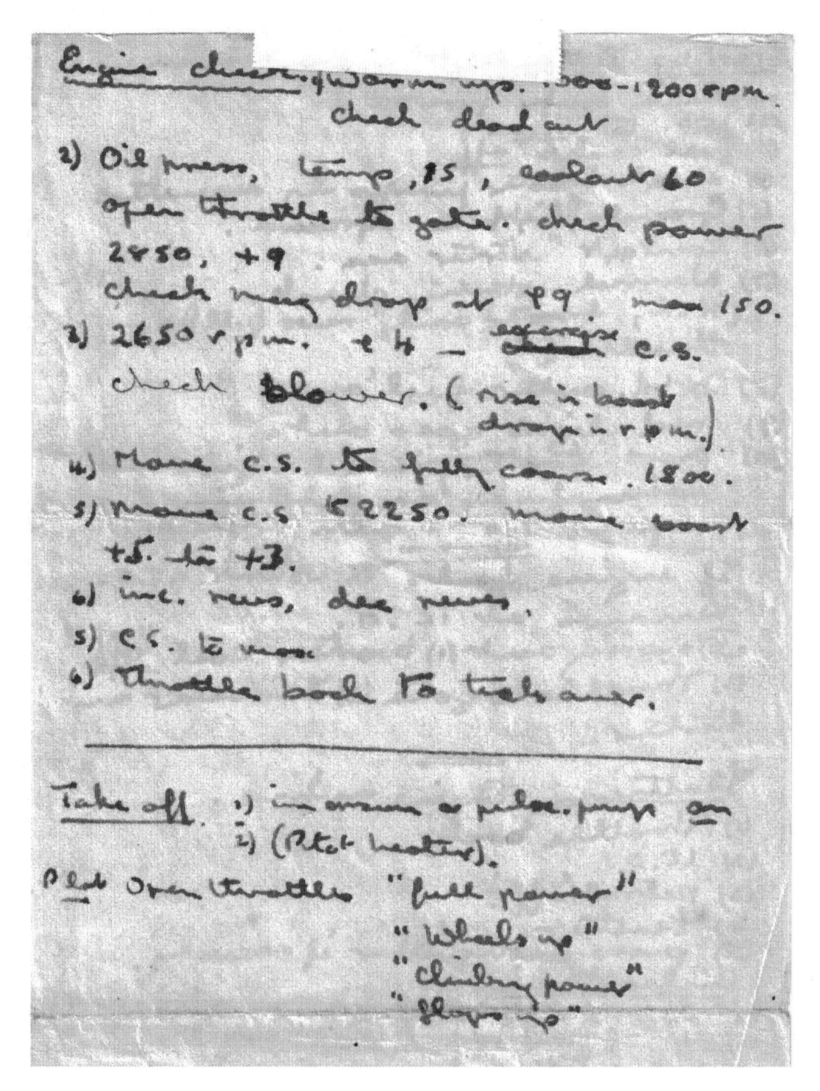
Engine check. Warm up. [illegible]–1200 rpm.
check dead cut
2) Oil press, temp, 15, coolant 60
open throttle to gate. check power
2850, +9
check mag drop at +9. max 150.
3) 2650 rpm. +4 — exercise c.s.
check blower. (rise in boost
drop in rpm.)
4.) Move c.s. to fully coarse. 1800.
5) move c.s to 2250. move boost
+5. to +3.
6) inc. revs, dec revs.
5) c.s. to max
6) throttle back to tick over.

Take off. 1) [illegible] props on
2) (Pitot heater).
Pilot Open throttle "full power"
"Wheels up"
"Climbing power"
"Flaps up"

the Germans would go and set up their own flares, which from a distance looked quite like ours. The German laid targets were always in a remote area just outside the town, in the hope that we would all fly over an empty paddock and bomb that rather than the town. I suppose that's another reason why it was so crucial to have a navigator, and to trust him, rather than just follow the crowd.

At this point, everything was still a secret, only those directly involved with the mission knew where we were headed, but that was only a slight advantage. The enemy by now had radar, and had learnt to filter out much of our false window foil signals. In fact, Martin Middlebrook, a British military historian, quoted in Hasting's *Bomber Command*, was of the opinion that whilst windowing had initially reduced the death toll of allied crew, by the winter of 1944 it had caused the Germans to focus so much on advancing their radar technology, that this period now suffered far greater losses as a direct result. There was nothing to stop the Germans detecting several hundred heavy bombers flying towards them. It may have been true that the allies were a little more advanced than the enemy in terms of radar technology, but the enemy now had very efficient detection systems, both on the ground and in the air. It was also because of their radar systems that we never flew in direct, straight lines. The theory was that zigzagging across the countryside would make it more difficult for the radar systems to fix onto aircraft and much more difficult to predict our target. Unfortunately I expect this only resulted in more hard work for the navigator to get it right.

In any event, it wasn't just the radar that gave us away. By now we had, through careful navigation and timing, merged with convoys from airfields all over Eastern England, to form a stream of about five or six hundred aircraft, so we were pretty noisy too! The enemy couldn't necessarily see us, but they could certainly hear us coming from a long way off.

The tactic was to concentrate on a specific target and hit it hard with many bombs. It sounded simple. It was not. Each crew knew where they had to go, and what time they had to be there, so they were all flying over the same target, just at slightly different times. There were usually two or three set bombing times about 20 minutes apart. This would amount to separate waves in which a few hundred aircraft would drop their bombs more or less simultaneously over the same target. We kept the briefing officer's instructions in mind; to be prompt in order to achieve the highest concentration possible, such that the enemy's defenses would be totally overwhelmed. From an individual perspective, this helped stack the odds. I am sure it was well intended as a comforting thought, but one of the blokes on the ground said to me, 'Stay with the pack, there can only be so many planes that they can fire on at once', I saw his sense, but I didn't like what it implied.

This meant that from the perspective of anyone on the ground, they would receive two or three concentrated 'dumpings', consisting of about 150-200 planes, each dropping 14,000 to 18,000 pounds of explosives, and all in about 20 minutes. This was the theory, but of course it didn't always happen so smoothly. People would tend to just get there when they could, and of course they weren't too keen to hang around, so they would often drop their bombs and quickly head off home again.

We flew in very large groups, a few hundred at a time, but not in tight formation—we couldn't see enough to fly too closely to one another, because we flew with no lights on at all. Without much moonlight and in total darkness we were thought to be safest, but it was difficult to see another aircraft even when they were right next to you. All we could do was to squint hard and try to make out the dark silhouette of a camouflaged aircraft against the night sky.

In fact, flying with no lights on, without radio communication, may have made it a little harder for the enemy to locate us, but it was a very real hazard. The idea was to avoid the German anti-aircraft weapons, while making sure that you didn't fly an erratic course that would cause you to crash into another allied aircraft. It was very difficult. Every now and then, one of the crew would look out the window and see the black outline of another aircraft wing, right next to us, and we would have to immediately adjust position, without bumping into any other aircraft. It was a good thing that it was too dark to be able to wave or see one another's faces framed in the little windows!

There were many mid-air accidents. Some pilots thought that by oscillating up and down frequently it would be more difficult for the Germans to lock onto them with radar. Others tended to think that weaving left and right may help them. I thought that was silly and decided to just fly straight and level. Apart from finding our way through the crowd, we had to navigate with total accuracy and maintain a course. If this wasn't

Propaganda Poster. Some sources claim this was used earlier to fool the Germans into believing the British had superior sight, not that they had more advanced radar systems!

enough of a challenge, we also had to avoid the enemy night fighter planes that would fire on us, as well as the flak they shot up from the ground.

All things considered, so far, our flight was going reasonably well. They tried to prescribe the heights during the briefings, but people naturally all crept up, and tried to get as high as they could in order to be less likely to be hit from anything that was shot up from the ground and to get out of the path of every one else. The other reason that climbing high was so attractive was to avoid the threat of having a load of bombs being dropped on your roof from some other friendly aircraft a few hundred feet above. So, from that perspective there were a few challenges. Given that the aircraft had a structural ceiling of 24,000 feet, there wasn't a particularly wide range and we couldn't afford to all gather at the exact upper height. Yet somehow, people generally sorted themselves out and ended up at a height and on a path that was not going to collide with their neighbour's. There were still quite a lot of mid-air crashes, but not as many as you might expect, or maybe it was just that there were so many other hazards that this wasn't our main fear.

We continued flying in the dark, heading east towards Berlin. Every now and then I would see batteries of ground fire, so we would have to avoid these. They were firing large anti-aircraft shells, and I hoped that the night sky would swallow me and provide some sort of invisible protection. There wasn't much we could do about it. The Germans seemed to fire at the large convoys, in an almost random fashion, hoping to hit one aircraft or another. They were good odds, for them. There were so many aircraft that some were bound to be hit. I don't know exactly how big these anti-aircraft shells were, I never stopped to look, but I do know that they could blow you out of the sky. It didn't matter where they hit the aircraft, it nearly always ended on the ground. Despite this, from some of the crash investigations, it was known that the Germans knew where the fuel tanks were and how to cause the most damage with a 20mm cannon shot, the Germans had done their homework too.

There were also fake explosions that were made to look like aircraft blowing up. The enemy would fire up a large metal shell, which would explode in mid-air—even if it didn't hit anything. It would look exactly like you would expect from fireworks coming out of a metal case. For a crew in the dark, it looked as though planes all around them were being shot at and exploding. It was supposed to have a psychological affect on the minds of the crews. There would be a large explosion, and every now and then bits of metal would tinkle down onto the wings of our aircraft, which wasn't nice. It reminded us of how vulnerable we were, but as we were already aware of the risks we were running, the spoofs and extra fear that these explosions were supposed to create didn't really make much difference; even fear itself seemed to have a limit.

We had learnt all about what to expect, but this was still my first flight. They had also told us that one of the things that the enemy was doing was to use a searchlight to highlight a plane before shooting it down. The enemy had a lot of very powerful searchlights, but only a few that were linked to their radar systems. There weren't many of these, but the ones they did have, would locate a particular plane, lock on to it via a radar beam, and then through their new radio controlled technology, the white searchlight would be turned to follow that plane precisely across the sky. Once the plane was in the spotlight, they would shoot at it, and cause a mid-air explosion. The clever part about this technique was the effect that it did have on the psyche of the other pilots. Normally, the enemy would shoot into the air and hit some planes, more or less at random. The difference here was that everyone could see the plane being highlighted and chased, darting about like a rabbit in the headlights. They tended to highlight it for a reasonable time before hitting it too, which of course built the tension. It really was one of their more effective psychological tactics.

Once we made it to Berlin we dropped bombs, hopefully on but perhaps only near, the target indicators. Gurus, the bomb aimer, called over the intercom, 'bombs away', in a voice that was supposed to sound excited, but did not disguise his fear. Some of the boys carried on for a minute or so, but their strained and out of character jokes revealed that they too had little to celebrate. I didn't feel anything positive for the successful delivery either, but I did breathe a sigh of relief; we were no longer a flying bomb. We still had to turn around and find our way home, but now the game was slightly different. Before long we would reach the predetermined place where we could use our radio, as there was no element of surprise left. However, they knew exactly where we were and would now be lining up for a second shot and although we were now rid of the amatol that had sat under our backsides, being shot out of the sky even at 20,000 feet wasn't without its problems either.

On this particular flight, our load had comprised of a large, 4000 pound bomb, six 500 pound bombs, and lots of incendiary—little 30 pound packages. I'm not sure the area that would explode under the force of a 4000 pound bomb, but it would make a big hole. It wasn't just the large bombs that did the damage though; it was the many thousands of incendiaries that were dropped on a town that would cause it to burn and after we had flown over, the town would often burn for days. The big bombs would wreck a few things and make life difficult for the Germans, but it was the fires that made the most damage and some of them really got out of control.

I didn't rejoice in this, far from it. I recalled hearing a speech by Churchill, in which he said, 'If they don't like it, they can retire to the hills and watch their home fires burning,' which may have been a clever use of words, but it made me shudder. They may have been the enemy, but they were still people. It was certainly not something that we wanted our leaders saying and it wasn't the way we felt about it either. I didn't mind bombing the military targets so much, that seemed fair enough, and a part of war, but being sent to bomb some places like Dresden, was just spiteful. A few of us talked about it amongst ourselves, but there was no place in the chain of command for our opinions on the selected targets.

We undoubtedly did a lot of damage, but in a city like Berlin, it would take many missions to make a huge impact, because although each plane carried a lot of explosives, and many incendiaries, so many landed off the target for various reasons. It depended on the pathfinders finding the target, and then the crews all finding the flares that the pathfinders had dropped. The system worked a bit, but it wasn't precise, so there would inevitably be a lot of collateral damage.

In fact, I think a lot of people would suggest that the greatest effect that Bomber Command had was the fear it struck in the hearts of the Germans and the way it boosted the morale of the Allies who had seen London, and other allied civilian cities, previously bombed. Bomber Command was often over estimated in her ability to do damage to the enemy. Indeed it could wreak havoc on a city, but more often than not, it would take many repetitive sorties to make any real difference and to slow the advance of the Germans in that town. They were a powerful force to halt, but for now, our superiors were happy that the Germans concentrated their efforts on the defensive rather than continue advancing on their other fronts. We felt the direct consequence of this, because they did have very good defences—enough to shoot down thousands of aircraft.

Chapter Twenty Eight
Touch Down

ON THIS FIRST SORTIE AT least, my crew and I had reached Berlin, delivered our bombs and then tried to make it home again. Those shooting up from the German soil and night fighter planes were just as aggressive as we tried to make our way out of the place.

As would happen on most return trips, the crews would all try to get home as swiftly as possible, and more or less arrive at the airfield at the same time, while it was still dark. As our navigator guided us in, our radio operator received a message to circle around at a specific height for a while. Eventually the control officers let us drop down to the next height and then the next, and finally join the landing circuit. Each airfield may have sent 20 to 40 aircraft for a particular night raid, so there was nearly always a queue. Depending on the weather and the condition of some Lancasters, it could take well over an hour to land all the aircraft in the dark with adequate spacing.

Coming home was always hazardous. Apart from being a weary bunch, there were always many dangers associated with landing. According to Hastings, the author of *RAF Bomber Command*, the death toll of 55,573 or 44.4% often quoted in relation to Bomber Command's losses does not include all those who died over UK territory whilst returning home. Based on a great deal of research, Hastings stated that an additional 15% would need to be added to the official death rate, to include those killed on their way back and I can well believe it, given the challenges of landing an aircraft before first light, and how many aircraft limped home, bearing great wounds.

While landing it was not uncommon for one aircraft to

run up the back of another, possibly in the holding pattern, or landing circuit or even on the ground. Being so dark, it was quite difficult to see the next plane, and thus, if they weren't spaced properly, a Lancaster could easily fly into the back of another. Sometimes it would only take the rear gunner out, but very often it would cause a multiple fatality.

Prior to my time at Elsham Wolds, there was one particular accident in which an entire crew burnt to death after a crash landing, not far from the airfield. One WAAF truck driver had been sent to the accident scene, and she had been given the bodies to take back. Their bodies had been laid out on the back of an open truck. As she was driving, the air rushing past the fibres of the airmen's uniforms had reignited their clothes and she had to suddenly pull over and deal with some bodies in the back which were now on fire. Another car soon came along and saw this poor woman, who was standing at the back of the truck, trying desperately to suffocate the flames and small embers by patting them down with her hands on the bodies. She hadn't anything else on her to use, and so she had done her best, but in the course of this she had burnt herself badly and was understandably distraught. The girls from the WAAF were competent in their own right, but as men, we still tried to protect them from anything like this, if we could, and when we couldn't we felt as though we had personally let them down. It wasn't something any woman should have to go through.

There were many other dangers involved with landing. In the case of those carrying the 22,000lb 'Grand Slam' bomb, it was decided, I assume by an official who resided in an office, that as these were very precious and costly bombs they were not to be 'wasted'. The implication of this decision was that if a Lancaster crew found themselves with any engine problems or a damaged aircraft, and needed to make a forced landing, they could not ditch the bomb at sea. Rather, they were instructed to 'carefully' land, with all 22,000lb of amatol under their seats. This was fortunately not a problem for me, nor for any of the men who were in my squadron during this time, but it made us feel rather dispensable.

On this, my first sortie, the landing went perfectly. We

taxied off the runway and began shutting down the engines. We had completed our first operation, and had only 29 left to complete a tour. When we came to a standstill and it was time to get out, I discovered that my back would not cooperate. I had sat in the same position for over 7 hours, on a freezing cold lump of a parachute pack. After a while I emerged a little sore, but it was nothing that anyone particularly commented on and it certainly wasn't anything that I wished to draw attention to. I had landed and for that I was grateful.

It wasn't until a few sorties later, that I would have my own landing problems. On 14th January 1944, I was told that my crew and I were to fly a brand new Lancaster. I had no clue that there should be anything suspect about it. Nonetheless, I did all my usual pre flight checks, which were lengthy and fairly detailed. I took off in the usual procession, and made sure to quickly clear off and make room for the long queue of Lancasters on the ground waiting to take off. A short while later, something just wasn't right with the aircraft, though it was hard to put my finger on it. We proceeded a while longer, but I still wasn't happy. Something was wrong with one of the engines. I decided to not go all the way across to Berlin, and instead I made the decision to drop the bombs over a much closer target in Holland, the nearest target, less than 90 minutes from base.

Once we were satisfied with our position, somewhere over Holland the bomber released the load. We saw the expected light show below as the bombs lit up the night sky, and then we quickly headed home. I felt satisfied that I had not wasted a mission, and knew that by now the stream of Lancasters flying in the dark would have finished taking off from our runway. With the aircraft the way it was, we needed to get back as soon as possible, but the condition of the aircraft must have deteriorated drastically because I soon decided that it was necessary to jettison the fuel. Unfortunately, this process didn't work properly, and fuel came out over the tank, into the wings and then flowed over the flaps, like a waterfall. There was petrol flowing everywhere as we were still going down the runway.

I thought this was not good, but it wasn't until my taxiing speed had slowed right down that I saw an airman dashing away, flapping his arms in a mad panic. What we hadn't realised was that it wasn't just the fuel the airman was desperately concerned about. When we had dropped the bombs, one of the flares had become stuck in the shoot. The flare had been dangling under the aircraft for the entire length of the runway!

It turned out that this new Lancaster had been manufactured in a hurry. The factory workers had gone on strike, and there were some oversights in their attempts to make up for lost time or lack of workers. It was an incredible set of events, to have had engine troubles, the fuel unable to jettison properly, and have a flare stuck. I think we all knew someone had looked after us that night!

The procedure after going on a sortie, particularly over Berlin, as I had done on those first missions, was always similar. We would return after many hours of gruelling concentration and no matter how much adrenaline was flowing around our systems, we would all feel exhausted. Unfortunately though, it was now the early hours of the morning and our work for the night was not yet finished. When we came into the main hall, we found the place was well and truly alive with busy workers, many of them WAAFs; the crew were not the only ones who had been up much of the night.

As we had been flying, those on the ground had been tracking the progress of each flight. While the aircraft were still outward bound they would initially be tracked on British radar systems. Further over enemy territory, cruising at their normal altitude of around 22,000 feet at 220 miles per hour, we had to rely on particular Lancasters with advanced technologies on board. These would transmit the wind speeds at their altitudes, back to base, and these would be calculated by the ground staff so that they would be able to accurately predict our timing and progress. From this part of the flight onwards, they had little to rely on other than our predicted and briefed course and timings, until part way home when we would reach the predetermined landmarks and be able to communicate with radio.

A short while later, the first of the Lancasters returning home would begin touching down on the runway. As each plane came home, it would be marked on a large chalkboard. A green round magnet next to an aircraft's name or number meant that it had returned home safely. One by one, the squares in the home column would fill up with green dots.

After they landed, each crew would make their way to the main hall. Our individual names were ticked off a long list. Next we were offered a cup of tea or coffee, each of which contained a dash, or sometimes, several dashes of rum. From there we would make our way around the room at various tables being

Bomber Command crews return, and enjoy a hot coffee with rum, before their debriefing.

interviewed by WAAFs about different aspects of our mission. We would be interviewed as a whole crew. They would ask strategically important questions for the planning of the coming night's sortie, such as, 'What were the enemies' defences?' 'What were the target indictor markings like?' and 'How concentrated was their flak?' At some point, it would be asked if we had seen any aircraft go down, and although this was always tricky to

'Interrogation' or debrief of the returned crew of an Avro Lancaster, 1943.

judge, by asking everyone the same questions, a broadly accurate picture was formed. From there, the positions of those presumed 'lost' aircraft were calculated as best as possible.

We would then move tables, clasping preciously in hand our warm mugs, and be questioned by those with a technical interest. We would be asked questions like, 'Did you have any problems with your on board cameras?' 'Were there any problems with any of your navigation or communication equipment?' and 'Were there any challenges with the armament or bomb release?' We were all very tired, and although I am sure we would rather have collapsed in bed, or be given time for our hearing to recover from the deafening sound of the Merlin engines, we did understand the importance of the debrief to be conducted so soon. They said information needed to be collated while it was still fresh in our memory, although how we were expected to forget it is beyond me, but in any case while we slept during the day, many of the dayshift workers would be using the data collected from the morning's interrogation for planning the next mission, that would be briefed to the crews going out again later that afternoon. They tried to form the best picture they could from what we told them, but we didn't always know how successful we had actually been. The real results came early in the morning when they would send the reconnaissance aircraft over to take photos in the daylight, and have them interpreted by the experts.

The interviews felt like an interrogation and were never popular, but what followed next was far worse.

CHAPTER TWENTY NINE
The Most Harrowing Part

AS SOON AS EACH CREW had finished with the panels in the interrogation room, they would go to breakfast. I suppose if one wanted to be truly optimistic, they could see the cooked bacon and egg breakfast as a huge compensation, given that by this time in England the rations were down to one egg a week, and bacon was even harder to come by. However, I always found the mornings in the mess to be the most dreadful part of an entire mission.

Slowly, after each crew had finished being interrogated, they would file in to the mess and sit down. It was the first time in the morning that we had all stopped moving about and were no longer in dispersed groups across the interrogation stalls. It was the first time that everyone gathered, and slowed down enough to look around. That was when we would see the empty seats. That was it. That was our debrief. There was no formal announcement of who had not returned, it was simply a case of eating breakfast while looking around the hall to see whose face was or was not there. Each time another crew entered, late from their interrogations, our hopes were raised, but inevitably, well after we had finished the meal, many seats remained empty. That was all. No one really said too much about it. It was something we all seemed to go through as individuals, and there was a real silence on the matter. It was as though it just happened without the need for anyone to make comment. For me, it was truly the most harrowing part.

The seats wouldn't remain empty for too long, as they were

always being filled with new crew, or 'sprogs' as we called them, and there seemed a near constant supply of young students ready to make their maiden flight. Still, for all of this, it was amazing that there was not a particularly heavy or depressed atmosphere. Somehow, we always found reason to hope; I suppose we had to.

The fact that our friends did not return did not always mean the worst had happened. Of course it was well and truly possible that these men had been killed in action. Sadly, as it would turn out, that was the case the majority of the time, but somehow, we always tended to remain positive. There were many alternative explanations as to why a crew may not have returned yet.

The most likely reason for any crew's late arrival would be that they had been diverted to another airfield. This did happen, particularly when the weather was bad, or if the plane had technical difficulties and needed to make an emergency landing. We could always hope that the message informing ground staff at Elsham Wolds of an aircraft's safe arrival at an alternate field was delayed. Then there would be optimistic suggestion about how it was possible that the aircraft could still be flying, and may have been delayed for other reasons, or have limped back slowly on a lean mixture to conserve fuel. These weren't unrealistic hopes, they were just not particularly likely.

Indeed, on 5th January, 1944, my crew and I had certainly put this theory of late arrivals and additional flying time on a lean mixture, to the test. It was my third sortie, and the fifth night that I had been at Elsham Wolds. I had been flying again to Berlin when we came under a new form of attack. It was such a new tactic that it had never before been mentioned in a brief. What the Germans did was pick out an aircraft from the stream of several hundred, and the chosen one just happened to be mine! Their radar then followed and got an exact fix on our location, all with its searchlight switched off, so at this point we didn't know about it. Meanwhile, the other searchlights were all hitched up to the main one, which was controlled by radar. Then suddenly all the lights were switched on at once. A great cone of light beams were all aiming up at a single plane—ours!

It was very dicey. We had been locked onto, and were lit up like a Christmas tree. It was really terrifying business, because the bright light beams made me temporarily blind. There was no doubt that we had been highlighted. Everything inside the plane was so bright. They were about to open fire on us, and cause a bit of a firework show for the others. It was clever of them. They didn't need the searchlights at all, they could have just connected the radar with the antiaircraft guns. The searchlights couldn't hurt us, but they certainly made us feel rather naked. The reason for the searchlights was to achieve two goals; firstly, they blinded the crew concerned which made for a panicked looking flight path, and secondly, by delaying the firing of the antiaircraft shells by a few seconds, they drew attention to the individual Lancaster, and highlighted its desperate struggles, which did have an effect on the morale of all the on looking Lancasters.

Caught in the lights, I darted about like a mad thing. I did everything to try to get out of the beam. I had started at around 20,000 feet, so initially I dashed about. Avoiding other camouflaged allied planes, alongside me, while blinded by the bright light was a bit tricky. I had hoped that they would have cleared as far away as they could have, if not to give me space, just to avoid being caught up in the likely explosion. I knew that dashing about had its hazards, so I put the nose straight down and then opened the throttle to get out of there as fast as I could. This had been over Stettin, so I realised that the nearest place to go to get out of the way was Sweden. It was so close, and it was neutral. I suppose this was a bit naughty. We really weren't supposed to do that, but after having just been the rabbit in the headlights, and very nearly shot down, it seemed like a good idea. This safe, neutral country stood out because unlike the blackened nightscape of Germany, the city below was all lit up with its normal street lights on. Once we had gotten out of the way of the German radar, they just picked another Lancaster and started the same coning process again. It wasn't nice! It was fortunate for us that this was a few weeks before they hitched up the whole system, the radar, searchlights and anti aircraft guns eventually became integrated so there was no chance of escaping the beam.

I flew over Sweden for a short while, and across the North Sea. This detour was making us late, so I knew that I would have to start conserving fuel. This part was interesting. With my vision restored and heart rate slowly settling, it was beautiful to see the northern lights, and the glowing atmospheric patterns. It was something new to see and we found it rather interesting, but most of all, it also gave us something to wonder at while we gathered out thoughts. We had been spared.

Then while still flying across the North Sea, we saw a small dinghy. We had no idea if there was anyone alive, but if there was a dinghy, then from that height we had to assume it was possible. Being just an isolated dinghy in the middle of the North Sea, we dropped height to see if there was anyone alive on board. The dinghy had a small canopy so we couldn't see anyone, but still we weren't sure. Meanwhile our radio operator called up Air Sea Rescue, and then we had to circle around the dinghy for a long time, so that they could get a fix on our position. For some reason, it took quite some time for them to do their thing, so by now we were very short of fuel. The flight engineer was very good, and he kept doing his calculations and letting me know how much more time we could possibly spend circling around. When it came time, and Air Sea Rescue still hadn't confirmed our position, he re-did his calculations, and we found a few more minutes. The count down was on. Finally, they confirmed that they knew our position, and we could leave. I then had to thin the fuel right out, and crawl back to England.

By the time we flew over the UK, it was approaching first light. We had flown almost 9hrs 30 minutes, on the fuel allocated for what should have been a 7 hour trip. We had so little fuel left that I expect the engine was running on the smell of the fumes. The engineer kept looking at the gauges and telling us that we didn't have much fuel. It was the first time that I had seen the airfield in the daylight. Anyway, we landed, and called up and said 'We've landed'. The voice on the other end of the radio said, 'Not here you haven't'. We were still rolling along the runway so I opened up the throttle and took off again. The runway we were looking for was only a few miles away. I had landed on the wrong field, because I hadn't recognised the place in daylight.

When we did land at our base, I taxied back towards the supply shed, and the first engine shut off on its own. The flight engineer said he wasn't surprised, as it didn't have any fuel left in it at all. By this time, everything was all shut up and the interrogation team had gone to bed. I had to call them up and get them to come out. I think they had almost written us off. Still, it did show, that even when people were not back for breakfast, there was some hope. There were some very happy mates who saw us later that afternoon.

Another thing that gave us hope, was the idea that a crew who had not returned, may not have actually died in an accident. Indeed, of the 125,000 aircrew, there was the well-quoted figure of 55,573 who died over enemy territory, but there were also a further 8,403 who had been wounded and 9,838 who became prisoners of war. Of course, we did not know these statistics at that stage of the war, but we did have the general sense that of all the planes shot down, there would have been many who did not die.

Apart from those who crashed on the way home, there were others who didn't return, but we wouldn't have a clue about what had happened to them. Sometimes we found out later, after the Red Cross managed to get a message back to the squadron or their families, but it took months.

Eventually news would make its way home, and there were many encouraging stories. Some of the airmen had bailed out and walked. It would have been a long walk, from Germany or wherever they were, but if they could get across into Switzerland or Spain they would be okay—although the latter would have required a long walk across German occupied France. Some people even managed to get across the North Sea to Sweden, but how they managed to do that, and take a boat, astonished me. Still, there were stories of people who walked for many, many miles and eventually arrived home safely, which was quite incredible. Fortunately this helped foster the ongoing feeling of hope, long after a plane had failed to return.

The RAF did give us some useful things in our flight kit to use in these cases, especially if we needed to walk anywhere. We were sent with spare rations, which would keep a man going for a while and little sweeties that the men kept in their pockets for emergencies. Apart from those who had to walk home, a great many of the uninjured crash victims became prisoners of war. The RAF also gave us some very neat pieces of equipment, which, if you could keep them on you and not have them confiscated when you were put into the prison, would be quite helpful. I suppose the chances were slight, but if it happened that you had bailed out and survived, became a POW and been fortunate enough not to have everything confiscated, then you were well set up for an escape. Overall, the idea was that you could then walk to a neutral country; it may have been Sweden, Switzerland, Spain or Portugal, which were all a long way away, but you could do it. People would sometimes get a friendly person to take them in a motor car, or if necessary, even steal a bicycle.

RAF issue 'Escape Button'.

I was issued some tunic buttons that unscrewed to reveal the face of a compass. This was a brilliant idea, but after a while, the Germans caught onto this, so they would check everyone's tunic. In response, the makers of the RAF uniform, put a left hand thread on the buttons, so that if any guard tried to check the button, then they would only be tightening it up and further concealing it. We also had trousers, with two buttons. If you unthreaded both buttons, and put one on top of the other, then located the area of the button with two little dots on it, that would be a compass. There were other things like a pocket comb, which when cracked open would reveal two magnets. If they were joined together by a piece of cotton they would also form a compass. Being so well disguised, in something as harmless as a comb or toothbrush, it was less likely to be confiscated.

People also took things with them that they thought might come in handy. One chap I knew had a piece of piano wire coiled up in his boot. Another man took the same sort of wire, but sewed it into the seam of his trousers. Then there were others with more brazen ideas who took things like a hacksaw blade with them, but they would usually get caught. The Germans weren't silly, but still, it was worth a try.

The RAF team who packed our flight kits included all sorts of other interesting things, in the hope that they wouldn't be confiscated. There was often a little part of a radio in most of the kits. The theory was that by being an indiscriminant little piece, it wouldn't be confiscated. Eventually, if there were enough allied POW's in a camp, between them, they would have all the necessary components to make a radio and transmit a message back to the UK. This did actually work and the POWs were able to transmit a signal. It didn't help with their immediate rescue, but it was very useful for the Allies' information officers. The prisoners would often see and hear little pieces of information that would be helpful, especially when put together in the context of a lot of other incidental clues, as I learnt later on.

Passport-sized photo of Cyril in civilian clothing, which he carried with him, in case he needed to obtain a forged identity card or passport.

Of course, the whole idea of being a prisoner, was to try to escape. The RAF had gone to a lot of trouble to teach us the methods that were most likely to help with this. For example, it wasn't too hard to find a guard who was not hired for his cleverness. From there, if you did it right, you could bribe him, but you had to be careful how you went about it. If the guard was a friendly fellow, people could ask for a toothbrush, or something that seemed harmless, but over time, if enough people managed to collect little pieces, they could make something useful to help

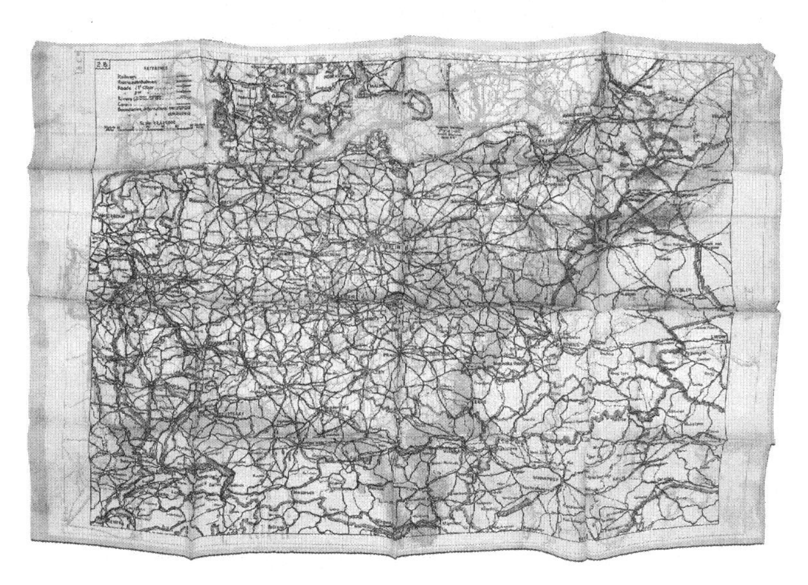

RAF issue silk map of Germany, beautifully screen printed on two sides.

them escape. It was also especially helpful if a POW caught the guards doing something they shouldn't, like smoking when they weren't supposed to, and then they could blackmail them. We also carried photos of ourselves wearing civilian clothes. The training officers told us that if you could escape, and then find a forger, they would be able to put your photo in it and make you a German identity card. We also used to take an odd scarf or handkerchief with us. It would just have a squiggly looking pattern on it, and could easily be disguised, but to the discerning eye, it was a map.

There was one chap who had bailed out and made his way across Germany. He came to give us a talk on what to do, and what not to do. He had compiled his experiences with other tips from other airmen, and taught us a lot about what did and did not work. They were mostly common sense things, but having them explained helped us as we may never have thought of them in the first place. He had been put on a train as a prisoner. He had asked to go to the toilet, which gave him the chance to move down the carriage, and so he seized the moment and jumped out as the train was going around a bend. Now, choosing the bend going in the correct direction was the most important part. If he had jumped out on the side which the train was bearing towards, then everybody could see him, especially the guards coming up in the later carriages. However, by jumping out on the other side, by the time the guards saw him, the train was already turning and beginning to obstruct their view.

The RAF also taught us places to go and people that we could trust. They explained that, generally speaking, you were safest with priests and prostitutes—not that they were often found together. They were always friendly and would look after you, no matter which side you were on. There were some interesting stories that came from this; there was an English bloke who landed somewhere and made his way to Paris. A French prostitute picked him up, and she looked after him, but business was business. She frequently entertained German Officers, but meanwhile he had nowhere to go, and it wasn't a well-furnished room, so he hid under the bed!

The priests were particularly hospitable. They would usually offer a well-earned meal, and a chance to freshen up, before putting an airman onto the escape route. With time, and the homecoming of many men, it became known that there was a well-established pathway. It was a completely unofficial chain of people, mostly related to the church, although not exclusively. It would usually start in France with a priest who knew someone who 'did this sort of work,' and eventually the airman would be passed along a series of people, each of whom would tell him the next person he was to go and see. It went from country to country, and multiple people had all made this same trip, but interestingly, for those in this chain, they only ever seemed to know the person directly before and after them, but no one else who was involved. The person would be passed from one helper to the next, who knew little else about the overall sequence or organisation. It was fascinating, and I would be interested to know how such a chain of helpers was formed.

The Red Cross also did their best to help, but they always remained neutral. They would distribute parcels to POW camps, and provide some medical aid where they could. The Red Cross were helpful, although they didn't take part in any exchange of information. They couldn't do it or they would jeopardise their position and limit their own abilities to help either side. However a POW could send a letter via the Red Cross, home to their mother to say that they were safe. This was done for nearly everyone, it was almost a standard process when one became a prisoner. The trick was to write home to Mummy, and if you were a smart cookie, you would include in there some information. It would need to be encoded in a subtle way that only someone in the know would identify it as a way of leaking a message. It would be addressed to a family member, who was already clued up, and, assuming the recipient picked up on it, they would forward it on to your relevant superiors.

On occasion you could also send a letter on to a superior via the Red Cross. In theory, the official letters would all be censored, however, there were such a lot of letters that inevitably some of them got through without being looked at. The Germans were well aware of all the little tricks that we would try, but with the number of prisoners trying to send as many letters as they did, some important information was successfully sent back.

As a part of our RAF education, we were told of a man who managed to get himself amongst the refuse and was picked up by the truck, while in a commercial bin. Then, when the truck came to a suitable place, he jumped off. The idea that there would be someone hiding in the rubbish would never have occurred to them. There were all sorts of surprising stories, people did the strangest things under war conditions and rather than being locked up, for who knows how long, people took terrible risks. Many upstanding young men in these situations were desperate enough to steal a motor car and to get a forged passport and cross the border. It's funny what war can do to people in desperation, and even stranger that stealing anything still felt so taboo, despite the absurdity of the reason they likely fell out of the sky and landed in Germany in the first place!

For the airman at base, it could take months to hear anything back from the Red Cross, or via any other channel about someone missing in action. This was difficult, because we couldn't finalise anything in our minds, and yet, it did allow for hope, which was vital for us, especially before we set out again that night or the next, on another sortie.

If you didn't come back after a sortie, within a few hours of your expected arrival time, you were posted as 'missing'. Then, without waiting much time, the officials would come and pack your kit and put it in storage or send it home to the family. You were just not there, and your bed was needed for the next poor fellow. There was very little way to know if people were okay, and about to walk home, or had died. So soon new crews would come and take over your bed and locker. For me, observing this constant turnover of people really was one of the hardest things of the war. It was such a cold and efficient process they had of packing up a man's belongings and making it seem as though he had never existed, it was truly harrowing.

The enormity of the situation always hit me in the mornings, when everyone went down for breakfast. It was strange, just a silly little thing like sitting down for a meal, but that was when it hurt. It wasn't the gruesome sights or the action packed moments that got to me. What hit me hardest was seeing a vacant row of seats in the mess, or a locker being emptied out. Of all my experiences so far in this war, that was the most distressing, and it happened nearly every day while I was there.

Most of the men reacted with blank faces and silence. They kept it all inside. I mean, what else could they do? It didn't seem that anything we could have done or said would have helped. The men would talk to each other a bit, but not a lot. I don't know, I am not a psychologist, but I would say that this wasn't healthy. When we came down to breakfast and saw the places where our friends used to be, we thought the best thing to do was to try to just carry on, and do our best to ignore it. As I said, we didn't know what else to do. It was very difficult, and this was one of those scenes that kept repeating itself. Every time we returned from another night flight and went down for breakfast there were people missing, marked only by the empty seats where our friends had sat yesterday. I suppose the officials didn't want to make it formal and do any proper announcements, for fear that they would make people despondent. Very often we would just go out and have a drink, before the next night. That was it. We just shut it off, for as long as we could.

My crew were supportive. We didn't talk about it much either, but we made allowances for each other, especially when someone a crew member was close to, or had trained with, went missing. We tried to distract one another, and keep things light, and Gurus, the bomb aimer, was marvellous for this. He was

Candy Gurus, in Cyril's sitting room in Wigan.

quite a bit older than most of us, probably in his late 30s. He was a most athletic man, and when things were down, or really quiet, he could just run into a room and do a flip; it was like a circus act. He would then come and pull out his trusty pack of cards, and demonstrate some tricks, and if you were really down, he showed you how they were done, which was generous of him. My radio operator was pretty quiet most of the time, but when we went out for a drink he knew how to let his stress out; one night we had to bail him out after he had taken a barometer off the pub wall and played it like a banjo.

Webby, our rear gunner was a good sport too. He used to help the navigator by calling out over the intercom every time we passed by a race course. Growing up in whatever part of Canada it was that had such a culture of race horses, whisky and guns, he was keen enough to be able to navigate across England based on the location of the race courses alone. He was however apt to show his stress, and became more and more particular about the condition of his guns. At first he would complain that they were not being cared for properly during the day while he slept, but then he solved this problem. He brought them in from the back of the Lancaster and decided to sleep with them under his bed! He was dissuaded of this after a few episodes and told that the men in the bunks next to him were not taking kindly to it.

After a few weeks in the job, we were granted three days leave, during which Gurus came home with me to my mother's and enjoyed a well-deserved break, with a few shots of whisky. Meanwhile, the mid-upper gunner, who had always been very quiet, took leave to go home to Dublin, and never returned. This probably rendered him the sanest of the lot of us!

Chapter Thirty
Lack of Moral Fibre

AN AWFUL LOT OF PEOPLE weren't actually missing; many were either killed in action or upon return. It was a terrible thing. With such high death rates, crews were constantly being replaced. The trouble was that new crews always thought they were fireproof. They had the idea that 'nothing could happen to us', but it would normally only take one or two sorties for them to settle; and a boy would become a man *overnight*.

After I had notched up a few sorties myself, and dodged a few bullets, I was acting as a Flight Commander. I had been sent to check out a new crew so I took them up on a few local circuits to test them out. They were enthusiastic and keen, yet somehow I could see what was going to happen. They were good at their job, but they were a bit too keen and eager. I told my boss that I thought this crew should be laid off for this first sortie, and introduced more gradually, or sent up with different crews on their first night. The new crew's gusto and enthusiasm was telling me something; it was as though they didn't understand what they were up against. I could tell they were just missing something of the big picture and I expected that they didn't fully grasp the risks. It wasn't as though they lacked skill or were incompetent, they were actually technically very good, but it was just a funny feeling I had.

I went to my boss and said 'I don't want them to go tonight, because I think something is wrong here'. I explained what I had observed and that I also didn't want them to go in the aircraft they were being sent in, the oldest and most worn out, simply because they were the last people to arrive, this morning. Everyone knew the particular Lancaster they were to fly was the 'knock about', and that it had limited altitude. Nonetheless, my superior informed me, 'My orders from on high call for maximum effort tonight, and this means everyone'. I explained to him again, that it was not the right thing to be sending them just yet, and suggested that they be split up and go with more experienced crews, but I was over-ruled on the spot. He said indignantly, 'They will go. There is to be no further discussion about it. They are all qualified, they can perform their tasks, so there is no reason not to send them'. I felt angry, and frustrated, but there was nothing I could do about it, there was a general flavour from on high that we were rather dispensable, and that if we lost a few thousand along the way, it was for the greater good. I wasn't about to turn the tide a few hours before a mission, based on a gut instinct.

It was a most tragic outcome; they didn't survive one mission. All seven of them died that night. It was awful, because I was responsible. I had sent them to their death. Seventy years after the war, this still haunts me and I often wonder if there was anything else I could have done to protect these keen young boys and delay their first mission. There were just so many factors in it that were wrong. That sort of thing happened far too often, and when I looked up the chain of command, the death tolls on both sides told the story of a lot of poor decisions. It was just the way it was, and we didn't question our superiors too often.

Surprisingly, despite all the heaviness of what we were dealing with, there remained a lot of optimism, for the most part. People tended to hold on to the hope of a fellow airman being 'missing' for as long as they could. It was odd how there always seemed to be hope, no matter how slight, until their death was confirmed. It was just one of those things; you did what you were told and you tried not to think about it too much. You couldn't afford to let it get the better of you, and the only way through was to cling onto the hope that you would somehow make the required 30 sorties. There was no room for fear or doubt. The reality of the situation did, however, get the better of some people, unfortunately. These people had problems with fear or grief, or both, or perhaps just the sheer stress. These were sad cases.

I imagine it was important for the RAF superiors that people didn't refuse to fly, even if they were struggling with the stress of the situation or fear of death. The RAF therefore had a system in place to discourage them from quitting or refusing. They would label a man 'LMF', if he decided that he couldn't fly another mission. This stood for 'Lack of Moral Fibre'. I thought it was a very unkind thing to say, but that was how it was looked at, especially by the superiors, many of whom had already completed a full tour and more. It wasn't of course a lack of moral fibre, it was just that they were scared to death. Nonetheless, if a man could no longer force himself to go, or be forced to go, the RAF would make life very difficult for him. Once labeled 'LMF', a man would be publically stripped of his rank, and then be posted to somewhere to do the most awful jobs. They were purposefully trying to make life for those who could not or would not fly, far worse than actually flying and facing death itself, so as you can imagine their treatment was pretty terrible.

I suppose the officials did have some understanding that people would eventually start to go 'round the bend' under prolonged stressful situations, so after you had done a full tour, they would tend to send you onto an administrative or senior role. Some people even got cushy jobs flying someone important to somewhere unimportant.

The problem with labeling men 'LMF' was that they made it very difficult for the men to admit when they were unable to continue and function safely. If they had not yet flown enough sorties, and their 'problem' was a mental one, their situation was not well tolerated. This resulted in an enormous amount of pressure on people not to quit, for fear of being treated as perverse 'LMF quitters'. As a result, there were a lot of men who were still flying, that really weren't able to cope, and many of them really weren't 'all there'; they had cracked under the pressure.

I remember one bloke, in particular, who was a skilled pilot, but then one day he just couldn't cope anymore. He had reached a point where it was all too much and no threat of punishment on the ground, matched his fear of flying or dying. They made his life hell. One of my close friends was a navigator and tried his best, but he just couldn't do it any longer. He 'downed tools' and froze in mid-air somewhere over Germany. He just stopped and could not seem to get his act together from that moment on. It must have been rough on the rest of the crew that night, but they did make it home.

This sort of thing happened all the time.

It was a very hard situation, because the men weren't really choosing to quit, it was just that their minds were failing them. They were trying to cope with too much. I think we did understand this to some degree, but the officials also had no choice but to make it a high penalty to discourage others from following suit. They created a culture where the men who could no longer go on sorties would be treated so badly that others would rather keep flying. The authorities must have felt they had to create and maintain some sort of culture, or there would be no one willing to do it. I don't know what the solution should have been. It was very difficult for everyone, and it really did mean that some people continued well past their maximum mental capacity, and as a result we saw some tragic cases.

There was one chap I knew quite well; he was British but had trained as a pilot in South America, where his family lived. During the war he had come across to England and joined Bomber Command. After a few weeks, he couldn't take the pressure of the situation, but he kept on going, which was, in my mind, a far worse thing than quitting. One of his crew told me that this pilot would get in the Lancaster, and the first thing he would do was light a cigarette. This was a real 'no-no', because of the sheer quantity of fuel and bombs on board, but he did it anyway. A concerned member of his crew informed me that this pilot sometimes took a flask of drink with him, too. I advised them and assisted them, as a crew, to all make an official complaint.

The Commanding Officer took the complaint seriously and said 'Ok, we will make the navigator the one in charge, and the pilot is just there to fly'. This was the only time I have ever heard of this happening, but it was inevitable what the outcome would be; the navigator could never have actually been in control over the pilot. The following day, his plane hit the deck and wiped the whole crew out.

I was sent to help investigate the accident site. I walked around what had been the Lancaster, which was now almost unrecognisable. I tried to separate myself from what I saw, and concentrated on the aspects of the situation that were normal, focusing on my boots as they crunched on the leaves below me. It was one of the most solemn and overwhelming moments. When I looked up there were bits of bodies hanging in trees. I desperately searched in the depth of my mind for a place to escape to; I wanted to shrink inside myself and hide.

Chapter Thirty One
Straining at Night

DURING THESE FLIGHTS, MY SACRAL and lumbar spine were not happy. Sitting had been my least favourite position since the crash on the beach in Pakistan two years earlier, and sitting still for seven or eight hours on a lumpy parachute, flying a thinly insulated metal tube through minus 40 degrees Celcius was not good therapy for a back injury either. My back wasn't particularly good when I first joined Bomber Command, but after a few weeks of these long flights, it got a bit worse. Still I felt that I was okay to keep flying. During the flights, I had so much else to keep my mind on, and watching others being shot down beside me, ensured that focusing on any pain was much lower on my list of priorities.

One of the things I noticed was that my injury limited my ability to move about the cabin at all. Mind you, when it came to moving about on the aircraft no one really had a great experience, so my crew told me I wasn't missing out on much. The only reason a crew member left their seat on these long flights was to use the portable toilet, or the Elsan as it was called, and this was only when it was absolutely necessary, as a break in concentration from anyone's task could endanger the entire crew. However, for eight to ten hour flights, I can understand the reason why they included an onboard toilet, and yes, I can vouch that it is near impossible to concentrate on life and death tasks in such cold temperatures, with the urgent need to take a leak. However, in any event, an onboard toilet really wasn't a practical invention.

The Elsan sat just a few feet forward from the rear gunner. It was rather exposed, and given the cramped space, lack of light, the need for oxygen masks and long tubes as well as the layers of winter clothes, the process was difficult. This was if all was going well. When in rough air, or if the skipper had to take sudden evasive action, things really got messy. As one unknown airman wrote, '*This devil's convenience often shared its contents with the floor of the aircraft, the walls, and ceiling and sometimes, a bit remained in the container itself*'. It certainly was not a popular device. Therefore, our crew, and a lot of others, would take little bottles and containers to use instead. It was quite common for such containers to be left somewhere over Germany, in what could be described as one of the early forms of chemical warfare.

It is claimed, that not all crews were as fortunate as ours when it came to nature's calls. The unknown pilot wrote, '*Returning from a mission over Germany, my crew had the trots and took it in turn to crap in a cardboard box which was quickly jettisoned. As the box fell, it hit the windscreen of a trailing B-17 in the formation and lodged frozen solid blocking the pilots view. Upon returning to base in England, the pilot had to land his aircraft by sticking his head out of the cockpit side window*'. So we were comparatively fortunate, and I was content to carry my little bottle, but even if I had wanted to, I don't think my back would have allowed me the freedom to use an Elsan mid-flight. Once I was sitting, there was no way I could climb out of my seat, and the emergency exit looked particularly far away.

When we would land, the men would typically all rush out, in part to relieve themselves, which often happened just there alongside the plane, but also to stretch and have a cigarette. This worked well for me, after the first few flights my back had become increasingly uncooperative by the end of each flight, so I had plenty of landings in which to develop a workable routine. I ensured that they all got off first, in part because it was a typical normal tradition for the pilot to leave last, but mostly because I didn't want them to see how much I was really struggling. It also took me that length of time to even get out of my seat. My crew all knew that I was injured, there was no hiding that, but I didn't want to draw attention to it.

I had learnt that with my legs being so jelly-like by the

An Australian Lancaster crew.

end of each flight, it was best for me to crawl over the mains power box, then hold on to the internal framework, and drag myself along towards the back door. It must have looked bloody silly. When I reached the ladder—which was needed because the Lancaster cabin was quite high—instead of climbing down, I would straddle it, hold on and slide down as slowly as I could. It worked all right, apart from the painful jolt at the bottom; but nobody seemed to object to this too much. I smiled and laughed and made a few jokes as we milled about at the bottom waiting to be picked up. By that time I usually had enough feeling back in my legs to let go of the ladder, which was a useful support, and walk around a bit more. This went on for quite some time, and nobody ever really commented.

Initially, I found that my back recovered after a short time on the ground. However, after a few weeks of distance flying, it was taking longer and longer for it to recover, and for me to regain strength and movement in my legs. It was a bit depressing, because I didn't know what was going on, but at least I could fly. I went eventually to see the Medical Officer, and although I can't claim that I told him exactly how bad it was, he did say that he had noticed that something was wrong. He had seen me shuffling into the mess hall on a number of occasions, but declared that as long as I was upright I could still fly. I wholeheartedly agreed.

Over the next few flights, I noticed that when I sat down, I would initially be comfortable, but then with time, my legs would more or less 'die' on me. There were a few occasions when my back really flared up and I was sent off to see various characters, but they never seemed to suggest too much. One time I went to the sick quarters, after losing both power and sensation in my legs, and they got me to lie on a hard surface for a few days, and eventually all seemed to go back to normal. I kept positive and after this time I actually thought my back was quite okay and on the mend. I suppose because I remained so optimistic, the doctors weren't too worried either, and they certainly didn't suggest I stop flying, so this gave me even further encouragement. Of course, there was no 'sitting down test'—if I had been made to demonstrate my ability to sit still for several hours in the cold, then I would have been declared unfit for anything. Nonetheless, I was able to continue flying.

During a couple of days of down time while in the sick quarters, my navigator, Woodfine, came over and thought he would chew my ear for a while. After learning that my father had been a ship's Captain, he returned ever so excited, with his sextant. There we were, at the airfield, and after all his careful planning and plotting, he informed me that we were in the middle of the North Sea, but closer to England than Norway, so that filled me with confidence!

Two or three nights later, on a night with almost no moonlight, I was asked if I would take a passenger, a man who was a technical boffin. He was a nice old stick, an elderly chap with silver hair. I was just asked to take this bloke and I said I would do what I was told, but I didn't really know what he was all about. He wanted to take all this radar equipment, to test the newest technology. That afternoon when we were delivered to our Lancaster, there was so much large bulky gear in the aircraft that I wondered how they found room to put any bombs in it. There were literally wires everywhere.

While flying a few miles short of Berlin, the elderly man would claim that he was detecting an aircraft to our left and then later to our right, and we would all peer out the window, until suddenly one of us confirmed that yes, they thought they could see a silhouette against the night sky. This happened a few times, and on one particular occasion, it was less than a half a wingspan away from ours. He was amazed how difficult it was to see an aircraft alongside us, we were more alarmed than amazed, and we certainly weren't amused. I wondered what else I must have missed.

After this, our radar expert, told us that he had detected an aircraft just in front of us, and that this was a twin engine aircraft, and said how many yards away it thought it was. He then asked if I wouldn't mind just getting a little bit closer to identify it and confirm what he had detected on his screen. Well, that request went down like the proverbial brick, but I obliged him anyway. I thrust forward a bit more, until we were close enough that even in the dark we could identify it. Yes, he had been correct, it was a German fighter plane! We quickly pulled back after that. Being a twin engine, I doubted that it would have had a rear gunner, well at least as far as I knew, it was unlikely, but it felt rather uncomfortable to be that close. We got through that all right, but by the time we landed the boys still had a lot to say about it!

Fortunately, by the end of this flight, my back was mostly cooperative, and while my crew talked to our passenger who was already on the ground, I managed to make my way out of the plane, without my creative exit being observed. In fact, normally I could manage okay by myself. I only recall one flight, a few weeks later, where I was completely unable to get myself out, however it worked out okay, in the end, because my crew just carried and pulled me out, across the mains power, down the back of the plane and towards the door. Then they all helped me slide down the ladder. Mercifully we were on our own at that stage, so there was not too much embarrassment and we had to wait a few minutes for the crew bus to pick us up from that part of the airfield. That gave me enough time to lean against the ladder and wait while I regained enough sensation and strength to be able to walk. I was able to carry on pretty much as normal once the bus arrived and do my best to blend in and not be noticed during the breakfast and morning interviews. The rum in the coffee was a good cover, or so I hoped.

After that incident, my back became increasingly painful, and I didn't have any painkillers. My problem was that I didn't want to press the point too much, because I didn't know what the M.O. would say, or if he would have stopped me from flying. I got away with it for a while, but the M.O. was keeping a general eye on me and he was becoming less and less happy with my situation.

Every month we had a Sunday service at the intersection of the two runways. We used to go there and do all the necessary things for a church service, which felt a bit odd, given what we usually used the runway for. Following this, we would have a march, where the Station Commander would stand near the flagpole. We marched along and did the smart salute as we passed him. That should have been fine, but this particular time I had to lead the flight crew through the march, and I started having real difficulties with my back. When I got to the flagpole, I did my proper salute, but in that very public moment, I lost focus on my feet and quickly fell down. There I was, laying on the ground, with the Station Commander looking down at me, and a few hundred onlookers.

I had saluted suitably, lifted a leg to march, and then promptly collapsed. I was so embarrassed, but the Station Commander thought this was hilarious. The M.O. was there, and came rushing up, he couldn't avoid addressing the situation any longer. I told the gathered help the first thing I could think of: that I had tripped on my shoelace. The M.O. whispered to me that this was not a good cover, because both my laces were still done up.

The M.O. prescribed more specialists and a lot of waiting. After a few weeks stay in the hospital I had been thoroughly immobilised but no one seemed to have much of an idea about how to fix it. My dear mother came to visit me during this time, and the nurses there were very nice, but overall, it wasn't a great situation. Despite the clues, I was still determined that it shouldn't be a major problem, after all, I had pulled through bad patches before. Webby and Gurus came to visit and informed me that I had been posted to Pathfinder, which was a real honour! This made me even more determined to get out of the stiff bed, but it wasn't long until the doctors conceded that they had no choice but to ground me. I was not allowed to return as a pilot with the Royal Air Force and worse than that, by 12th April 1944, I was posted to a world of paperwork.

At the time I was greatly disappointed. However, years later, I see that it was probably for the best. After I was grounded, the crew that I had been flying with were split up and asked to fill vacancies with various other crews. Occasionally as the war went on, I would get news of how some of them were doing. Sadly, out of the original seven of us, four died in the following year. I had been spared that, but it came at the cost of a lifetime of debilitating pain.

It wasn't until nearly fifty years after the war that I would see a surgeon who could help. Since the landing on the beach in Pakistan, my back had always been problematic, but I hadn't understood the significance of the incident until this surgeon explained it to me. When the nose of the aircraft had gone into the soft sand, I had slid forward and lifted off the metal seat. As the plane bounced back down, I connected with the edge of the metal seat and fractured my sacral and lumbar vertebrae in multiple places. Initially, I knew it was very sore, but I didn't realise quite how much damage had been done, and at the time I needed to get on with being rescued. Once Rennie and I got back to Egypt, I continued to have a lot of pain, but I thought it was due to the bruising and would heal. With time, I justified it as muscle aches because it roughly coincided with my frequent malaria relapses.

Two years after that accident, when the doctors X-rayed it, the fractures in my vertebrae had in fact 'healed'—just not in line with how they were initially designed. I was never given a copy of my medical reports, but was informed by a rather busy doctor, that there was, 'a small kink towards the bottom', but not terribly much to worry about because 'the bones had now settled and were firmly fixed in their new position'. In 1944, as

EFB.

(4) R copy (3)

MINISTRY OF PENSIONS,
SANCTUARY BUILDINGS,
18, GREAT SMITH STREET,
LONDON, S.W.1.

In any further correspondence on this subject please quote :—

Telephone : Abbey 1200.
Telegraphic Address :—
Warpension, Parl., London.

0/M7/20238.

1st May ~~April~~, 1947.

Dear Sir,

In reply to your letter dated the 18th April, 1947, I am to refer to the Notice of Decision sent to you on the 10th April, 1947, in which you were informed that you had a statutory right of appeal to the Pensions Appeal Tribunal on two issues, one of which concerns the finality of the award, and to state that if you disagree with the Minister's decision it is open to you to exercise the right of appeal or not as you so desire.

The degree of disablement arising from sciatica is assessed at 6 to 14 per cent. temporary less than a year. I am to explain that this means that in the Minister's opinion your disablement is assessed at 6 to 14 per cent. and the effects of war service will disappear during the ensuing year.

I am

F/Lt. C.A.B. Johnson,
Holly Villa,
Haigh,
Wigan,
Lancs.

Letter from the Ministry of Pensions, 1st May 1947.

I received this news, I realised there wasn't much that could be done, so I didn't press for further details.

By 1947, when I was struggling with severe back pain, my family doctor wrote to the Ministry of Defence Medical Office. The Ministry of Pensions wrote back, unambiguously! They informed me, that not only did I not qualify for a pension, but concluded that 'the effects of the war service will disappear during the ensuring years'. I accepted this, as I understood there were many other people who were more injured than I was, and there weren't enough resources and medical services to provide for those who should have been at the head of the queue.

Almost a decade later, while I was still struggling with the 'temporary' sciatica, I was formally assessed with regards to the degree of the damage. This time it was found to be 20% but no solution was offered. Many years after that, I was reassessed, and found to have a 30% disablement arising from the crushed spinal nerves, but by then I had made my own way in life, and they weren't offering any surgery either, so it was of little consequence. It was not until I was 70 years old, in 1990, that an Australian doctor would offer a surgery that had a significant chance of reducing my pain. I wrote to the relevant UK authorities and asked for assistance in covering the cost of the operation, but by the time the appropriate correspondences had passed, a few years had elapsed. In mid 1993 I wrote and advised that I was not impressed with the length of time that it was taking them to identify the obvious, and that I thought it was a bureaucratic disgrace to 'keep veterans waiting until they died out'. This did not particularly help my cause, but it did help with my frustration! At glacial pace, a few more letters were exchanged, perhaps sent in a corked bottle from the UK to Australia.

By March 1994, I still had not received any useful correspondence, apart from a few odd letters, typically advising me to write to another person in the same department, with a different name. The following is some of my response to the ludicrous situation:

"It is not my intention to appeal against the decision of the Directorate. I am not prepared to fight the umpire and I have neither the desire nor the ability to do so... I am fortunate that

my wife looks after me by doing those things with which I have difficulties… Should anything change what should I do? Write to the department? I would be long dead before I got a reply; I am still waiting for a reply to my letter of June 1993… One of the biggest irritations in dealing with any Government department is the total anonymity of the letter writers. Who, in your department, next to God, makes the decisions? You can tell me, I am so far away in Australia, and struggling to stand straight, so he need not fear I will thump him. Every time you write to my address, you get ME!"

This did not particularly help me either, but it does sum up the absurdity of the longstanding situation.

In any case, in 1993, I elected to have the surgery. The neurosurgeon advised that it should not be a particularly difficult or lengthy operation, but even he was amazed at what he found. After eight hours of surgery, a few bone grafts, two plates and several screws, he informed my son that he had found a fragment of splintered bone, which had lodged itself in some precarious place, and that at any stage over the previous 50 years, with a bit of a jolt, I could have severed a nerve. So that was indeed fortunate.

Chapter Thirty Two
Intelligence, Secrets and International Relations

HAVING BEEN GROUNDED ON ACCOUNT of my lower back on 12th April 1944, I was posted to Bomber Command headquarters where I was offered a new role, as an Intelligence Officer. I suppose they thought the most useful thing to do with a man who struggled to sit, was to put him somewhere behind a desk all day—military logic—a concept that still defies me.

The training course itself was held in London. It was well run, and I learnt a lot of things I had never heard of, or even thought about before. The course only went for a few weeks, but they crammed so much in during that time. It was designed specifically for Air Force Intelligence Officers, and there were only ten students, so we could ask a lot of questions. It was fascinating. Naturally, we weren't allowed to speak about these things, but I suppose that nearly seventy years on, nobody cares too much if I share a few secrets now.

When I had been in various flying schools, climbing the ropes to Bomber Command, I had often been told never to carry anything on my person that may identify who I was, or where I had come from. This included little incidentals like a train ticket or a bus pass that may help the enemy to piece together where the airfield was located.

It was not surprising to learn that the Allies also played the same sorts of games.

In fact, I soon found that some of the British intelligence people weren't particularly nice and did some nasty things; to say they were a sneaky bunch would be an understatement.

I remember learning that there had been a young German airman who had been shot down over the UK. He was captured and had done something fundamentally wrong; he was carrying letters from his lady friend in his tunic pocket. From this they managed to identify who he was and where he came from. That in itself, may not have been too bad, but unfortunately for him, the RAF also discovered that he was married. The young man wanted the letters from his mistress back, and he certainly was not keen to have these love letters forwarded on to his wife. So, he paid for these letters with information. Indeed, based on the instructor's account, I think he paid for the letters several times over with the amount of information he was required to give. According to those who were running the intelligence course, this was quite legitimate and all very fair, at least in their eyes anyway. They asked him all sorts of questions, which he should never have answered, but under the circumstances he spilt the information.

Those running the course shared an incredible amount about what was going on so that we were even more equipped to ask the right questions. As a part of the course I was sent to Bletchley Park for a few days; which was fascinating indeed, albeit, one of the strangest places I have ever visited.

Bletchley Park was a Victorian Gothic style mansion acquired by the British Government at the beginning of the Second World War, to house the Government Code and Cypher School. Here they sent some of the greatest intellects, many of them geniuses in mathematics, languages, engineering and specialty fields, including cryptology and even papyrology. There were even world champion chess players amongst their number. It was probably the most eclectic bunch of brains that had ever been gathered!

Their job was to crack the German Enigma codes so that the enemy's messages could be intercepted, and the Allies could respond accordingly, an achievement that would shorten the war and save many lives. My job was to just observe, an opportunity for which I was grateful. Thousands of messages would come in, and the workers, operating in three eight-hour shifts, would enter the codes from the encrypted messages, searching for patterns. They were working on the Colossus, the first computer prototype, which was the size of a room. It was able to handle multiple data inputs at once, although it did require a lot of effort and frequent servicing to keep it operational. Then someone tried to give me a brief outline on how a Turing machine worked. Most of the data

Colossus, the first computer prototype, Bletchley Park, England, 1943.

processing that I observed was well beyond my comprehension, but not beyond my appreciation.

At Bletchley Park, there were many people who you could identify instantly as being academics from one prestigious place or another, as they wore their unmistakable comic like professor glasses and white coats, but then there were others who you wouldn't know exactly what they did or how gifted they were until someone else pointed it out. It was also the place at which I observed women being more highly regarded and respected than anywhere else. I was initially surprised how many women there were, and when I asked my room mate about this, he said, that in 1939 when it all began, some thought it was best to keep them out of it all together because women were much more likely to gossip, but they did allow a few in and it had worked out so well that now they did the lion's share of the work. Given the way our government painted women in the propaganda posters, in which it was suggested that it was the girls who talked and whose "loose lips sank ships", they were certainly trusted and well respected here! Many of the ladies were also particularly fluent in multiple languages, often being educated at foreign boarding schools, and they were credited with a lot of the success stories. There were all sorts of people here, but strangest of all, was the complete mixture of age, occupation, class and gender, and it all worked so well!

Bletchley Park was a fascinating place to stay, and there were some of the strangest people I ever met there too! The man with whom I was sharing a room had been there a few years, and he told me some incredible stories about his colleagues. I didn't know what to make of those, but in some regards, it felt like a bit of an alternate universe. Nevertheless, from what I saw, they all had a good lifestyle and routine, and there was very little sense of segregation or even class. People all ate in the same cafeteria, with no table distinctions, and during their breaks they used to all go outside and play rounders, including the professors!

By the end of my course, I had learnt much about what was really going on in the war, and I understood much more about how the enemy worked, and the strategies of both the Germans and the Japanese. The thing that struck me about the Japanese was that they had come from the Middle Ages type conditions, to a real first-class nation within 50 years. They were incredibly good at copying technology and thus, as a nation, they developed rapidly. Having been catapulted into the modern world meant that they had advanced military technologies, with a mindset for battle that was more or less from the Middle Ages. Hence they played by very different rules. They would chop a man's head off without issue. They thought nothing of the most barbaric actions, many of which may have been a part of European history several centuries earlier, but with all their modern weapons, the actual outcomes were in fact far, far worse. I don't think they thought that life was particularly sacred, well certainly not somebody else's life.

The overall idea, now that we had finished our course as Intelligence Officers, was that we would conduct interviews, designed to help pick up little bits of information and put them together like the pieces of a jigsaw puzzle. We were taught what to look for, and how to specifically dig deep with our questions around areas of information that would seem totally irrelevant to most people. Once we learnt how to put all this information together, we soon found that many things, which would have seemed trivial, were in fact vital clues. We also learnt a lot of very peculiar and fascinating things that people would do both in their public service lives and also in their private lives.

I learnt how to interrogate the enemy, however I did not enjoy doing this at all. Fortunately, a big part of the course focused on teaching us how to interview our own men and piece together information, which on its own, the men may have thought to be irrelevant and not bothered to report. It was all about anticipating what someone may have seen, and asking the right sort of open-ended questions. I tended to focus, and show great enthusiasm, for learning how to interview those from the Allied side, and interviewing our own people became my specialty. We formed connections from the little clues that we gathered during our countless interviews, and when we put these facts together, with the bigger picture that we had formed for ourselves, we discovered a great many secrets.

Mind you, I suspect we sometimes jumped to the wrong conclusions.

After the intelligence course had finished, I was posted to an airfield at Prestwick, Southwest Scotland. I think this was because I was a Flight Lieutenant and there was a need to have someone slightly more senior to oversee the place. Of course, I didn't really know too much about what was going on. It must have been necessary, at least on paper, to have someone with a senior rank running the show. When I arrived, I was amused to learn that my assistant was in civilian life, a fully qualified solicitor. He knew a lot more about what I was supposed to do than I did.

Prestwick was a very large airfield, on the main air corridor for all the aircraft that flew from North America to the UK.

Above: Prestwick Airfield, and Orangefield house, Southwest Scotland.

Below: Rear of Orangefield house with control tower.

Almost everything that came across the Atlantic touched down, to at least refuel, in Prestwick.

My job was to interview those who were returning from operational or transit flights, particularly over the north Atlantic or enemy territories.

From time to time, an allied officer would report that he thought he saw what looked like a U-boat surfacing somewhere out in the ocean. To a pilot reporting this, it may have seemed inconsequential. It was well known that most of the enemy U-boats had to surface for fresh air at various points in their journey, as even with a snorkel they still tended to do this periodically. Sooner or later, one of the team would interview somebody else with a similar sighting at a slightly different time and location. By plotting these sightings from various people, we were able to predict the speed and course of many German U-boats. At least that was the theory, whether we got it right or not I'll never know. These facts were all collated and used by the Navy, but we rarely heard any more.

I also had to look after security in regards to things going out of the country. For example, anything flying off to a neutral country like Sweden may have information on board that was being sent to Germany, although we probably only suspected this because of the games and tactics that our own side had tried during the war. Such information could have been disguised in the form of letters, or forwarded by word-of-mouth from the crew, or anything else, so we would have to be quite careful. Still, there was the need for discretion in how we intervened, especially given that the flight was outbound to a neutral country. We couldn't control them, we could only try to minimise the risks. On occasion, the continual trickle of leaked information was used to our advantage, and someone high up would think out some scheme and progressively feed bits of information through the known leaks, to confuse the enemy and distract them from what the Allies were really planning.

Before my assistant had come to Prestwick he had been posted in Northern Ireland. It was here that he honed his incredible skills for making friends with the Americans. He was a mild and meek sort of chap, with all the typical solicitor characteristics that one might expect. He was careful and thorough about everything he did. There had apparently been something suspicious about the Yanks, so he used his skills of negotiating and building friendships and after a time, he had formed such close bonds with them, and such a high degree of trust, that they had given him, an Englishman, access to their internal filing cabinets. When he got the opportunity, he read what was in their files. He copied information and sent it across to our headquarters. I'm not sure what those at headquarters did with the information, but he had confirmed our suspicion that we really couldn't trust the Yanks.

The solicitor shared more about what he had read; whilst they were fighting the war in Europe, the Americans were simultaneously planning the opening of a large international airline business straight after the war. They were planning on taking over the major flying corridors they had been using to cross the UK and many other regions of the northern hemisphere. The Americans had been going through files, personnel records and all sorts of other sources to find out everything they needed for the opening and operation of this proposed airline. They knew that the airline would be highly profitable, and they were making plans right down to the level of detail that included who to poach or hire after the war. So much for being our gallant allies.

One day, while I was working in my office, the phone rang and I thought it would be one of my friendly colleagues. It was not. Normally, at this time of day my friend who was at the other side of the airfield rang. As I picked up the telephone I answered in a rather official sounding voice, 'This is your Commanding Officer here' and the bloke on the other end just replied, 'That's funny, I thought I was'. I had been caught out. Fortunately he had a good sense of humor and took it all rather well.

From time to time my Commanding Officer gave me some very odd jobs. This particular time he had phoned to ask me to look after a Russian Colonel. I was surprised that my prank answering comment hadn't deterred him, but when I realised he was serious about this request I apologised, and, trying to make it sound sincere, I said, 'I do not speak Russian, so I don't think

I can help you'. Unfortunately this excuse didn't really work because, as I had expected, the Commanding Officer informed me that the Russian Colonel spoke perfect English.

When I was eventually introduced to the Colonel, I was amazed. I had always thought that the Russians, being Communist, would be a rather strange lot and difficult to get along with, but actually he was one of the most well-educated people I've ever met and we had plenty to talk about. Culturally the experience was very interesting, because the Colonel didn't understand why we did the things that we did, things that at that stage, people in the UK considered perfectly normal and never questioned. I took him for a walk along the seaside to see the town and experience the town and port of Ayr, near Prestwick, Scotland.

After a while, we came across a small arcade and found some game machines inside. They were early model pinball machines and children would often come along and put pennies in the slot to buy themselves a game. On this particular occasion, when we entered the shop we were the only customers. The Colonel looked at the game machine and asked, 'What are these?' I talked to the attendant and said, 'Give this bloke a handful of pennies' and given the impressive stature and decorations on the Colonel's uniform, this was, 'No trouble at all,' for the young shopkeeper. I put a penny in the slot and began to show him how they worked. I think the particular game he was playing was, 'What the Butler saw' although there were quite a variety for him to choose from.

It had been a quiet day in the shop, until two little children walked in and one of them put a penny in a slot. I could not believe what happened next. The Russian Colonel stood up, put one hand to his hip, stomped the other foot and bellowed angrily, 'For chiiildrrrrrren'. He promptly marched out. He was most offended that I would think that he would enjoy something that was for children. Of course, they weren't all for children, far from it in some cases, but he couldn't see that. After that he was thoroughly unamused which was a pity because until then he had actually seemed to be really enjoying the game. It was the hallmark of pride.

We continued on our walk as though nothing had happened. Not a word was spoken about the matter, although I did notice his posture was now even straighter and he strutted with his shoulders sitting even broader than before. The Colonel then said that he fancied a drink. I explained, 'I'm very sorry but it is Sunday and in Scotland this means that all of the shops, and especially the pubs, are shut'. He replied in a rather overbearing tone, 'But in Moscow, at any time of any day, I can always buy a drink'. I thought to myself, I can't just let this go, I need to try and see if I can help him, after all it was my job to entertain the man, and not giving a Russian Colonel a drink when he demanded one, just didn't seem right. I was only a junior compared to this big shot, so when he asked me for something, I felt it necessary to think of a strategy that would fulfill his wishes, no matter how silly.

With the Colonel in toe, I went and found the residence of the proprietor of the biggest hotel in Ayr, knocked on his door, woke him from his afternoon nap and asked him, would he please open up the hotel. This proprietor was not so sympathetic towards my desperate desire to please the Colonel, and even if he had been willing to open up the pub, the Russian was by now very unimpressed.

Given that I couldn't satisfy his desire for a drink, I proposed a quiet walk along the promenade. Halfway along our walk, we came across three or four people from the Salvation Army, who were singing hymns and playing an old accordion and a group of spectators had gathered to watch. I still remember exactly what the Russian said, 'Look for children' so I said, 'Yes, that's right' and hastened my pace, being careful not to intimate that I thought he could be interested.

I had intended to keep walking past, but he stopped and stared at them awhile. I tried to edge onwards, and keep him walking, but by now he had noticed that they were wearing their white and navy uniform with various epaulettes. The Russian Colonel asked, 'What army is this?' Foreseeing this conversation was unlikely to go well, I was keen to avoid the question so I repeated, 'Yes, for children', hoping this would dispel his curiosity. It did no such thing. The Colonel pressed the point until I said

'This is the Salvation Army'. He asked, 'Salvation, what is that?' and looked at me frowning, but as soon as I got to the word 'church' in my explanation, he shut off completely and marched away with quite some speed. He didn't want to know a thing about it from then on.

He was a very polished man, incredibly well-educated, but so terribly arrogant. He was completely removed from the realities of the world. The perfect Russian. I don't suppose one would get to the top brass without those characteristics, not in the Russian military. I had been selected to escort him because there was no one else, but I didn't have any training in this area. I hadn't done a very good job of the whole day. This certainly wasn't my cup of tea. I was only a little cog in the system. I suppose there was no one else around who had a reasonably appropriate rank, so that would have been the only reason I was given the role. I was used to playing with airplanes—that was my thing—not entertaining the top brass from Russia.

I couldn't tell you if I preferred dealing with the Russians or the Americans. Both seemed as arrogant as each other but I think, in the end, you could know the Russians more easily than the Americans, who were never really that upfront about their intentions. Mind you, these two nationalities didn't get on too well with each other either, even at that time, when they were still all supposed allies. They didn't trust each other at all.

After being stationed at Prestwick a few weeks, the Americans announced they were going to sell some of their aircraft to the Russians. Therefore, several Russian Air Force crews had been sent across to the United States to look at their aircraft before the deal was signed. The deal must have been a success because the American aircraft were now being flown by Russian pilots to their new base in Russia and they were coming via Prestwick for a brief fuel stop. It was my job to arrange for the refueling of all these aircraft and I was supposed to then send them promptly on their way. The task should have been simple, but it was not.

The Russians landed their aircraft and wanted to refuel and take off again almost straight away. One man asked how long it would take to refuel, and a member of the ground crew said, 'We can have you refueled and off again in less than half an hour'. The Russians didn't seem to believe this answer so they challenged him, which led to the grounds man replying, 'I'll bet you five quid we can'. It was the sort of casual one line throwaway remark you'd expect from a young worker who was confident with his abilities and knew his job well.

The next minute these Russians were in my office insisting that I should arrest this worker for trying to 'bribe a Russian official'. The Russians then told me that what they wanted me to do was to take him out the back and shoot him! The Russians wanted to get the military police involved in what they viewed as a case of corruption and bribery. The poor man from the ground crew had only been trying to be friendly. I took this young fellow out the back and hid him in one of the hangars. I told him to stay still and not to come out for quite some time. I returned to the Russians and told them rather expressionlessly, 'Yes, he has been dealt with'. They were well satisfied with this, and soon after someone else had refuelled them—without the friendly banter—they were on their way. I didn't know what else to do, but fortunately the day ended well and everybody seemed happy.

By now, I had formed my own opinions on how the Russians operated. With regards to the Americans, my assistant and I shared the view that the Yanks really were all profit orientated. These opinions of mine were well and truly confirmed after the February 1945 Yalta Conference.

The Yalta Conference, sometimes known as the Crimea Conference, was held in Russia. It was a meeting held between the 'Big Three' heads of government, Winston Churchill, Franklin Roosevelt and Joseph Stalin. Indeed, these 'Big Three' allies were incredibly powerful forces at this time, and although they only met together like this twice during World War II, it was without question that when they met, the course of world history changed.

The three leaders of England, the United States and Russia had gathered together, knowing that an Allied victory over Germany was almost assured at this point in the war. Essentially, they were gathering, before the war was won, to carve up the

The 'Big Three' at the Yalta Conference,
from left to right, Churchill, Roosevelt, Stalin.

spoils. Of course, this agenda is worded far more diplomatically in most history books, but essentially, that was what this conference was about.

All three governments came to the meeting with slightly different agendas. Whilst a victory over Germany seemed inevitable, the Pacific theatre still was in doubt. Japan was fighting strong, and thus England and the United States saw a great strategic advantage in establishing the guaranteed participation of the Soviet Union. The conditions under which Stalin would agree were discussed. The Soviets were chiefly interested in obtaining more land and strengthening their conquests. The primary concern for Britain was to maintain her vast Empire. The Americans were seeking assurance that the Soviets would join the United Nations, which was in its infancy and, above all, the Americans and the British wanted to know that the Soviets would join them in the war against Japan.

All three nations agreed that the Soviet's reward would be a sphere of influence in Manchuria, (an extensive region of North East Asia) following Japan's surrender. I am not sure if anybody thought to ask the occupants of Manchuria, but nonetheless, Churchill and Roosevelt were willing to give it away as a participation gift. Apparently this was the major accomplishment of the meeting between the three gentlemen. All who were at the meeting seemed to achieve their own ends, so they must have been rather pleased with themselves at this point.

Yet, still the conference pressed on to achieve more. The future governance of Poland, and then Germany, were topics also on the agenda. It was frankly and openly negotiated as to who would have power over which areas, as the men carved up great chunks of Europe between them. Discussions were also held as to any allotments for France and the appropriateness of de Gaulle to attend any future meetings like this. However, Churchill said, in his best humour, that this was a, 'Very exclusive club, the entrance fee being at least five million soldiers or the equivalent'. Indeed it certainly was a cozy little meeting they were having.

Oddly enough, I knew all about this meeting, but not from any of the controlled propaganda that the media released. The public were informed of the grandness of the conference, and enlightened by articles from a perspective that highlighted the power and might that was represented at such a gathering. The media told of the strength and grandeur of the three men in an attempt to build up confidence and patriotism. There were images of the Heads of States, strutting in to the Lavadia Palace and being saluted as they filed past very senior officials. It was all supposed to remind us of just how powerful and strong the allied governments were and the way they would rule the world. However, I formed a very different impression.

Prestwick Airport was the first port of call on the long flight home for the notable officials and scores of people who escorted them. Their aircraft all stopped for refueling and some other services before continuing on the long journey home. Somehow, after this, an entire copy of all the notes from the Yalta Conference ended up in my filing cabinet and under my care. Why all these notes should have been kept in my filing cabinet

I have no idea. I suppose, it must have been something to do with my being an Intelligence Officer, but I didn't see anything intelligent about leaving those documents with me. I was told that I was to guard these top-secret documents and keep them under lock and key. I was certainly not given permission to read them. Of course, after a little while, I read them anyway. I'd have been silly if I hadn't done so. I justified that, otherwise they would just have been making use of me and my filing cabinet, and I didn't particularly like that.

When I read the Yalta Conference notes I found them rather dull. I was far more interested in what the blokes that were flying across the Atlantic and over enemy territory were saying and what was happening with the actual war and the real people who were out there risking their lives. I wasn't at all excited about this political nonsense and the idea of dividing up the spoils of the war, before we had even won it, especially not while my friends were still out there fighting. I was very unimpressed. After that, I decided not to lock the cabinet, and to just focus on interviewing the airmen, learning where the actual hazards were, and the things that were likely to make the difference between life and death for our men.

After a while, I then reconsidered my move to leave the filing cabinet unlocked. This sort of information, although I didn't like it at all, was clearly not for public eyes. Besides, it was cluttering up my cabinet. I got my Woman's Auxiliary Air Forces officer, or 'WAAF' who was a wonderfully loyal secretary and went with her to the filing cabinet. I opened the drawer and grabbed the great wad of documents, and marched out of the office with her following closely. We went around the corner, where I stopped and screwed some papers up, and dumped the rest in a loose pile on top. I then set fire to it all. I am not sure exactly what the 'WAAF' must have thought.

It happened that my Commanding Officer had come to visit me that day, which was actually a rather rare occurrence. While we were standing around our little bonfire, he marched around the corner, presumably attracted by the smoke. There wasn't much I could do. It was pretty obvious that he was going to ask some questions, so I stood to attention and informed him that I was burning the documents from the Yalta Conference. He looked at me a while. The silence felt awkward. I went on to say that I didn't feel that they were very secure in my filing cabinet. He continued just looking, glancing between me and the smoldering pile of ashes. He then said, 'Yes, but there is a procedure for closing files'. There was a small pause and then he smiled a little and congratulated me for using my initiative and walked on as though nothing had happened. I never heard a thing about it after that.

They were rather interesting times I had at Prestwick. I suppose there was a great deal of responsibility that I carried with the nature of my role, but with that came the trust and respect that enabled me to do what I thought was right, even if it wasn't always exactly in line with official procedures. I didn't feel particularly watched or checked on which was a good thing, because it enabled me to just get on with the many and varied tasks I had to do.

Chapter Thirty Three
Boarding with Mrs Clarke

I ENJOYED MY TIME AT Prestwick and particularly admired the beautiful Scottish countryside. In those days, Prestwick didn't have an army barracks, or anything like that, so we were billeted with civilian families, although the exact details about how all these sorts of arrangements were made are beyond me. I think that anyone around the area who had a spare bed was either asked, or told, to take someone in. From the first day of my posting here, I was billeted with a lady, Mrs Clarke, who was in fact a widow with two daughters.

Mrs Clarke had lost her husband after the First World War, from the wounds he sustained when a sniper shot him multiple times. Her two daughters were both away at war. The youngest, Moira, was in the Auxiliary Territorial Service, the branch of the army that was open for women to join, and she was based in the south of England, working in signals. She used to help pick up information from the German broadcasts. The eldest daughter was a theatre sister at a hospital in Glasgow, which accounted for the spare bedroom and why Mrs Clarke had volunteered to host someone. It was a very modest, but comfortable little home and she always made a great meal, despite all the rationing.

Mrs Clarke was a lovely lady, about my mother's age, and we got on rather well. Whenever she would go shopping, I would go with her and stand in the ration queue to enable her to go off on some other short errand. It helped her, and it made me feel useful too. It felt good that I could also help out a little bit around the house with a few odd jobs. One day she had a barometer that

Mrs Clarke and her two daughters, Elizabeth (left) and Moira, 1943.

Mrs Clarke's youngest daughter, Moira, typing messages.

needed mending, so I did this with a piece of red cotton, and it still works nearly seventy years later. We really did get on very well, and had a lot of amiable conversations over many cups of tea.

Mrs Clarke sitting with some of her patients.

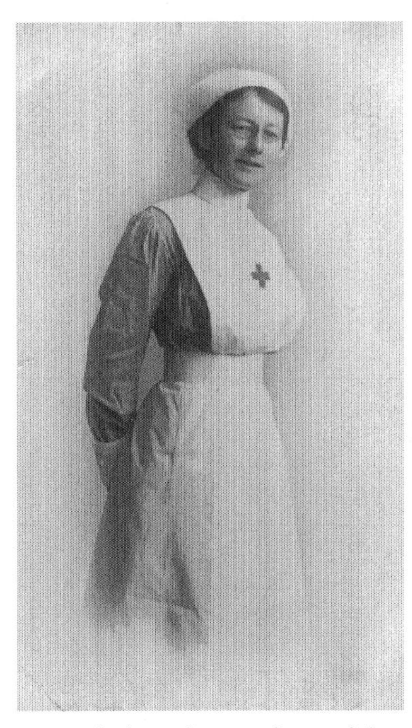

Mrs Clarke in her uniform of the Voluntary Aid Detachment during World War I.

Mrs Clarke had been a 'VAD', from the Voluntary Aid Detachment in World War I. The VAD was formed from the Red Cross and St John's organisations, so she initially worked in the ambulance service. Of the 74,000 VADs in 1914, two out of three were women who were employed for their first aid skills, as well as their nursing abilities. Others were trained as cooks, hospital orderlies, gardeners and carers of varying levels. After Mrs Clarke had spent some time working on the ambulances, she was transferred into the hospital and trained as a nurse. She spent a great deal of time with her patients, and that was how she met her husband.

Mr Clarke had been badly wounded. The Clarke family story said that he was a sniper who had served in the army for many years before WWI started. One day, during the war, he was unable to find his normal uniform pants, so he wore his second pair, which were a shade lighter. This may have been fine on other nights, but it was a full moon, and with his pants being a bit brighter, he was seen by an enemy sniper. He was shot multiple times. Fortunately, he was wearing a locket watch, and it was when I heard this story that my mind cast back to Africa and the superstition that a locket would offer protection. In this case it did! Some shrapnel hit it, and did great damage to the clock, but it prevented him from being killed outright. He was also shot in the leg, and consequently spent a long time in hospital. He recovered sufficiently to marry and raise two young daughters, dying a few years after the birth of their second child. Mrs Clarke explained that he died from an infection due to his

Mrs Clarke, furthest lady on the right, in the days when nurses were also expected to tend the gardens, c.1916.

wounds sustained in that action, but she didn't share any further details.

One day, after I had settled in, and felt at home at Mrs Clarke's, I arrived home to see a strange lady standing in the living room. Unsure what to say beyond the usual niceties, I went through to the kitchen where Mrs Clarke was and whispered, 'Who is the young lass out there?' I was then introduced to the eldest daughter of the house, Miss Elizabeth Clarke. She was a beautiful lady. Elizabeth had come home for a few days leave. While I may have been fond of her from the moment I first saw her, she was actually rather upset to have returned home and discovered that someone had been sleeping in her bed. She was not impressed with me at all!

Mr Clarke on Crutches after he was wounded during World War I.

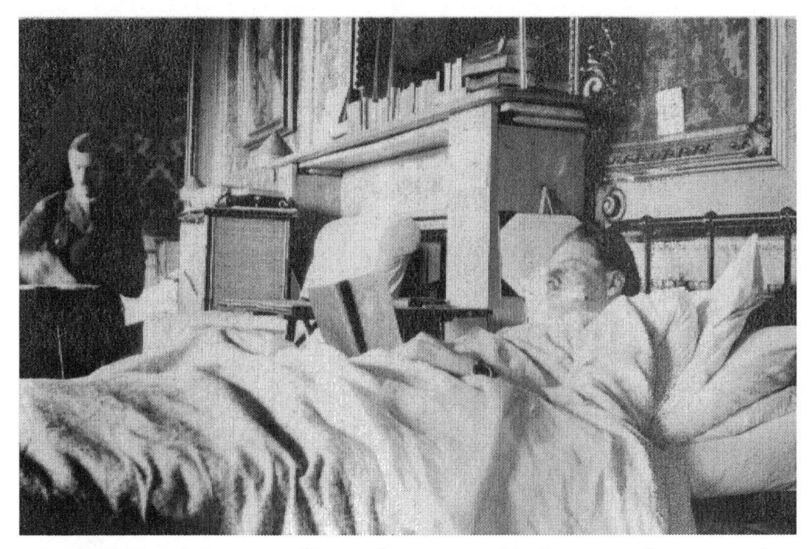

Mr Clarke in his home, suffering from wound infections a few years after the end of World War I.

In those days, the old-fashioned Scottish houses had a 'Lit Cloe', a sort of cupboard, usually in a living room, which would open up and reveal a mattress base. This mattress could then be folded down and made into a temporary guest bed. Elizabeth had come home from working hard in the hospital and was unhappy to find that she was going to have to sleep, more or less, in the cupboard. Immediately, a rather good solution sprang to mind, but it would have been improper even to joke about such things.

Elizabeth had been living in Glasgow, in the nurse's quarters of the hospital. It was only a thirty minute train journey away, but nonetheless, she usually came home for one weekend every couple of months. At first I didn't pay too much attention to her, so I certainly didn't encourage the frequency of her home visits, but after a while we became rather good friends and suddenly she started to return to her mother's place nearly every weekend, which her mother and I were both very pleased about, although possibly for different reasons.

Elizabeth, or Bet as I now called her, had trained as a theatre nurse, but what exactly theatre nurses did I didn't quite

Elizabeth, 'Bet' Clarke, in her nursing Quarters in Glasgow.

know. We never spoke much of our roles in the war, but I know she saw a lot of terrible things. I knew the sorts of horrors I had seen bound for hospital and the numerous war injuries that clearly required surgery, and so I imagine that much of it would not have been pleasant. Still, it was strange how whenever we got together, we managed to leave all those things behind. It was as though we didn't have a care in the world.

Bet and I spent a lot of time going out around the town and surrounding villages, sight-seeing. We sometimes visited the movie pictures. Other times, we would visit her three aunts who lived in a magnificent house, with very large gardens on the top of a cliff looking over the ocean, called, 'Nethermuir'. All three

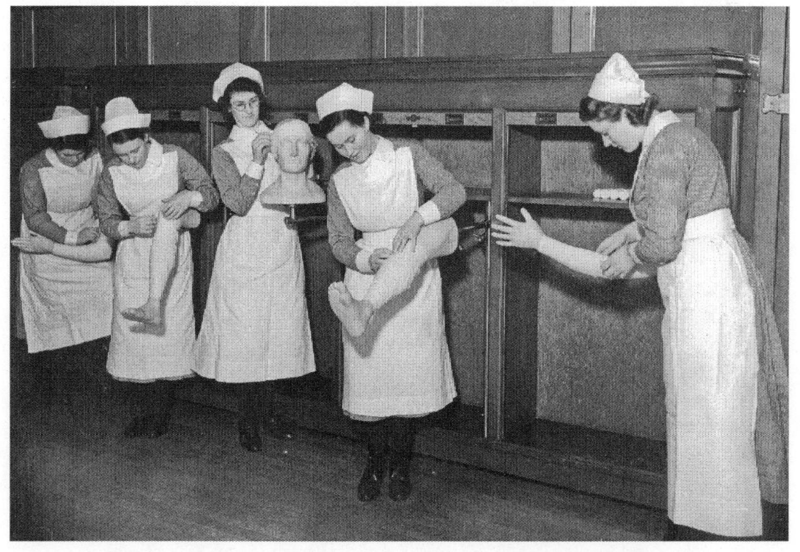

Elizabeth training as a nurse, second nurse from the left.

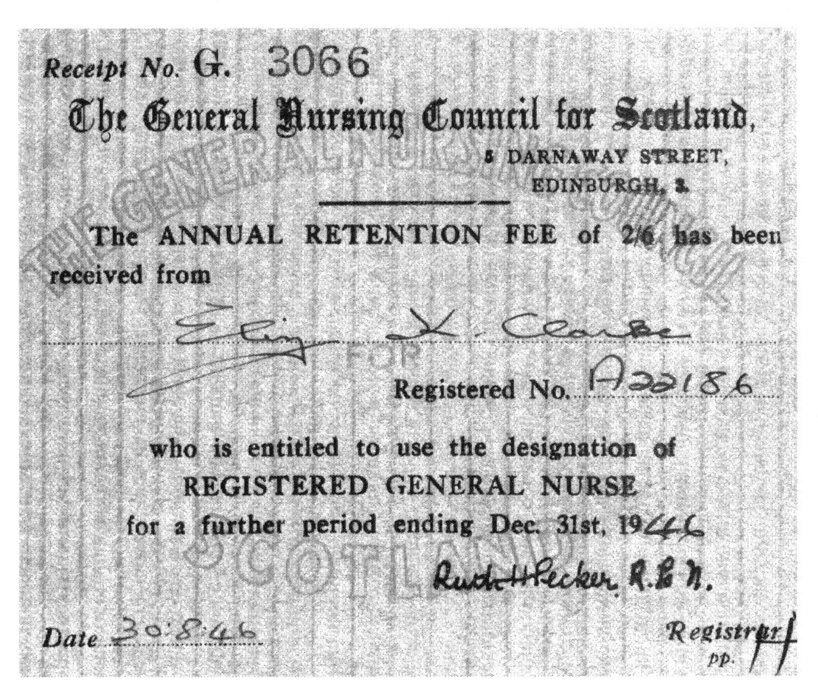
Receipt No. G. 3066

The General Nursing Council for Scotland,
5 DARNAWAY STREET,
EDINBURGH, 2.

The ANNUAL RETENTION FEE of 2/6 has been received from

Eliz. K. Clarke

Registered No. A22186

who is entitled to use the designation of
REGISTERED GENERAL NURSE
for a further period ending Dec. 31st, 1946

Ruth H Pecker. R.G.N.

Date 30.8.46

Registrar pp.

Nursing certificate, and receipt of annual registration fee.

aunts had all lost their men in World War I. In the period before the Second World War, they used to entertain and do charity functions, and whatever else it was that people of high society did all day. My grandparents would have approved. Of all the many famous and interesting guests they entertained, the most remarkable was Helen Keller. This young lady was both deaf and blind, and yet became a world famous political advocate for those with disabilities. She also published a dozen books and several articles. She was truly inspirational.

When we weren't visiting family, we would often go for walks along the beach, or sometimes along the river, Bonnie Doon, and quote 'Ye banks and braes oh Bonnie Doon'. It was beautiful. I had a very old bicycle and Elizabeth had an even older one, but they did work. Hers was the very upright sort, often described as a 'sit up and beg' type of bicycle, with a little cane basket on the front and for some reason, the bicycle was named Percy. Over the back wheel were special cords running

The old Clarke family home, 'Nethermuir'
where Elizabeth's aunts now lived.

down to prevent a long skirt from getting caught in the spokes. She and I would go all over the countryside on picnics and enjoy our time together admiring such a beautiful part of Scotland with green rolling hills, perfect for an afternoon lunch.

Not far from Mrs Clarke's house, was the birth place of the poet Robert Burns. The poet's house had been preserved, and we both enjoyed visiting. It was a two-storey stone villa, built in the mid 1700's, with a large stable block. It looked wonderfully quaint and peaceful, with holly growing all the way up the walls. Around the home, was a well kept garden with maples and ever greens and even some apple trees. It was like something from a children's fairytale.

There were all sorts of places to explore on our bikes. There was a river on the way to Burns' house with a little cobbled stone, humped back bridge and a great place nearby for picnics. Mind you, we didn't have much to have picnics with, but still, we managed. Our dating activities were quite limited, and it was a close knit community so we couldn't get up to any trouble, but we did get to really know each other by having picnics and talking for many hours at a time.

Elizabeth on her bicycle, in front of 'Nethermuir' house.

It was a beautiful historic town. In fact, Burns used to visit the local pub at Ayrshire, which was several hundred years old, and he had inscribed some of his verses on the walls.

'O, saw ye bonnie Lesley,
As she gaed o'er the Border?
She's gane, like Alexander,
To spread her conquests farther!
"To see her is to love her,
And love but her for ever;
For Nature made her what she is,
And never made anither!"

By now I had known Elizabeth, or Bet, for six months, but we had only seen each other on most weekends for the last two or three months. She wasn't like any other lady I had ever met. She was so easy to get along with; we didn't need to ask too many questions, we just understood each other so naturally. We were both quite pleased with the situation.

I found a suitable time and went and asked Mrs Clarke if I could marry her daughter. She said, 'Well, that would be up to her, but I certainly don't object to it,' and I had come to learn that she wasn't the type of mother to interfere anyway. A couple of days later, Bet and I went for a walk along the River Doon. I couldn't help myself a moment longer, and asked, 'Will you marry me after the war?' She said, 'Yes, of course, what took you so long?' I then realised the enormity of what I was suggesting.

At that stage, I didn't even have a ring, but I quickly remedied the situation. I couldn't afford much at the time, but a week or so later I gave her a ring; it wasn't exactly out of a Christmas cracker, but it may as well have been. It was more of a token gesture, with a promise that I would get her a suitable engagement ring just as soon as I could. We then went on a river cruise, I remember it so clearly.

Shortly after that on 8th May 1945, Germany surrendered and the war was over, well at least it was over in Europe, but instead of being sent home, I was sent to the Far East, to help in the fight against Japan. I was instructed to go to Burma, now

Myanmar, as the Group Intelligence Officer. I knew very little about the Japanese, but I soon learnt a lot as I was stationed there for about nine months. It would have felt like a long time just being in Burma, normally, but with my fiancée on the other side of the world, and having to delay our wedding, it felt like a an eternity.

Bet on a river cruise when we were first engaged.

As soon as I had saved enough money, while in Burma, I went to find a local jeweller, who was rather interesting. He told me that when the Japanese had invaded, the Burmese had buried their valuables. By the time I went to see him, it was late 1945. I walked into the shop and explained that I would like to have an engagement ring made. He then produced some Victorian Sovereigns, old English coins, which were made of 24 carat gold. He explained that he had just recently dug them up again, now that the Japanese were no longer a threat. We then came up with a design and I picked out three ruby stones. I was so excited to be able to present it to Bet, just as soon as I could get back to the UK. I missed her terribly.

While I was in Burma, Bet and I used to write to one another as often as possible. It was the highlight of my week to receive a letter from her in the post. One letter I clearly remember; there had been an outbreak of chicken pox in the hospital. This was in the days when it wasn't known that the most infectious time for spreading chicken pox was before the spots came out. Bet was very concerned that she may infect me through her letter, so she wrote it all out and got her mother to put it in the oven, to kill the bugs. I received a letter, and the envelope looked normal. I opened it up and the letter fell to pieces in my hands. The paper was toasty and brittle. I had to stick it all back together like a puzzle before I could read it.

Sadly, we didn't keep the boxful of letters. We made the decision to throw them out at some stage many years after the war. It was just one of those things; life changed a lot as the years went on and they were thrown out. It is a great pity now, as I would very much like to read them again. Fortunately, we did keep all the photographs.

Chapter Thirty Four

Stepping into the South-East Asian Theatre – After the Show

WHEN I WAS FIRST POSTED to Rangoon, Burma, I arrived to find that for some reason, I had been put in charge of a lot of people. I had been asked to take over the role of a Squadron Leader, and told that this was my new rank, although my Wing Commander advised that being so far away from the UK, I shouldn't expect the official paperwork to catch up with my promotion for quite some time (now, nearly seventy years later, I am not expecting it in the mail any time soon!). This posting may have seemed a natural progression for an Intelligence Officer, but I still saw myself as a pilot, albeit with his wings clipped, and I definitely had no expertise in this field. Mercifully, these people seemed to know what they were doing so that was all right. The chap that I was replacing was much more senior and experienced, but he had retired. I hadn't the slightest idea what I was to do, but I soon picked it up. It seemed an odd choice to send me, it wasn't really in line with my experiences, but nonetheless, between my team and I, we soon worked out what it was that I was supposed to do.

My time in Burma was interesting. I was given several roles and tasks, each of which brought about their own challenges. This was particularly so by 2nd September 1945 when the Japanese, who had been occupying the place and running the country, surrendered and the Allies were left in charge. It was all over, bar the shouting, well that and the re-organising of a few million people. The lives of Burma's 60 million residents had been completely upended, and there was much that needed doing before they could regain even a semblance of normality. There was an incredible amount of chaos and the Burmese local civil authorities were by now almost non-existent. Finding anybody who was in charge, or who knew what or who they were in charge of, was difficult. On top of that, there were some rather hostile attitudes that had been brewing.

To make sense of the situation, it was necessary for me to take a step back and look at some of the key events that had occurred in Burma throughout the duration of the war, but even then, things were not always clear. I had wonderful staff, many of whom had served in this theatre throughout the war. They did a great job of bringing me up to speed, but it would not be until decades later, the 1973-4 British television documentary series, *The World at War,* would help fill in the blanks, such that I could understand more of the history that I was a part of. With this insight, I can now look back on my experiences, and understand the events that were part of the underlying causes for the confusion and mess that I walked into.

Unfortunately for the Burmese, the location of their beautiful nation set them up as centre stage, hosting some of the lengthiest battles and retreats of the entire Second World War. To Britain, this nation formed a natural barrier that would protect its Indian Empire. For the Japanese it was a new source of rice and vast quantities of oil. To the Chinese it was of great strategic importance, as the last remaining overland supply line. No one could afford to loose control of Burma.

In the opening years of the Second World War, this was a very one-sided theatre with the Japanese being particularly experienced at jungle fighting. They seemed to do well under these conditions and made the natural environment work to their advantage. The British on the other hand, took a while to work out that they could exchange their Yorkshire pudding for bamboo shoots and reduce their need to carry endless cans of Bully Beef. The British and many of the Allies also struggled to adapt to this hot tropical climate. The majority of Burma is a jungle, and to the British soldiers it felt like a steam sauna most

Allied troops taking a rest in the Burmese jungle,
showing something of the incredibly lush vegetation.

of the time, and even in their summer uniform, they were never free of the perils of the jungle—having to be careful of the ongoing threat posed by mosquitoes.

The monsoon rains in Burma were unbelievable. Where I was in Rangoon, or as the locals called it, 'Yangon', there would be rainfall averages of more than 20 inches per month. Even more surprisingly, much of the country's coastal regions received an annual rainfall of 196 inches. It was wet: It was real wet! If one imagines the heaviest rain possible in England, going on for several weeks at a time without even a day's break, this would be monsoon season in Burma. The roads tended to all wash away and everything turned into a thick sticky mess. As one veteran would later describe it in an interview, for the television series, *The World at War*, '*Squashing through mud, living through mud, lying in mud, sleeping in mud, drinking in mud and eating in mud. That was a monsoon in Burma, it was just a nightmare*'. The monsoon season lasted for about six months of the year, from May to October, with warm temperatures all year round, ensuring the most incredible growth of leafy green vegetation. It seemed as though almost half of the country was comprised of steep mountainous terrain and covered completely by thick closed canopy jungle. Apart from the transportation and physical challenges for a land army living and fighting in these conditions, such a climate also promoted all sorts of diseases.

Health ailments and diseases plagued both sides, but particularly the British, who were initially ill-equipped to deal with such conditions. Diseases such as scrub typhus, dengue fever, prickly heat, dysentery and malaria were rampant. Further, the men with their reduced supplies soon suffered from jaundice, beri beri and jungle sores. I don't think many of the Allies had ever even seen a jungle before being freshly deployed to Burma, and even in our fairy tales and wildest imaginations; there were none so thick and unrelenting as this. It took the men a little while to figure out how to cope in such an environment, and to learn the ways of jungle living, including the art of burning leaches off their skin with a cigarette lighter, preventing skin diseases when walking in wet boots all day and other skills that would eventually become second nature.

British and Allied troops in Burma, 1945.

Meanwhile, the Japanese were doing well; not only were they familiar with the environment and climate but they were also used to dealing with such health issues. From all accounts, they had actually been thriving, using the jungle as cover as well as a source of food. For the Allies, I suspect the jungle felt like just as much of an enemy, as the people they fought, so many of them found their time here incredibly oppressive.

In December 1941, the Japanese had invaded Burma, with the advantage of surprise, in jungle based hand to hand conflict for which they had the all the advantages. The results were pretty good, if you happened to be Japanese! Not so for the Allies, or for that matter, the poor Burmese. In an interview broadcast as part of the television *The World at War* series, General Sir John George Smyth, a Victorian Cross recipient and former Commander of the Indian 17th Infantry Division, said, "*I don't think any country was more unprepared for war than Burma was at this particular time. The government was unprepared, and the civil organisation and the people were unprepared and the defence forces practically didn't exist.*" He went on to explain that when they were first invaded, he hadn't a wireless set with which he could even contact the air support in Rangoon. Therefore, when the Japanese advanced, he exclaimed, '*Believe it or not, the only thing I could do was to tap in onto the railway telephone line, get [the worker] in the Post Office in Rangoon, and try and persuade him that it was vitally important for me to be put onto Air Force headquarters. And that was really one of the reasons why in our withdrawal to the Satang, we were terribly badly bombed by the RAF as well as the Japanese Air force*'.

Poster produced by the Americans, reminding the Allies of the threat that mosquitoes and malaria posed.

Allied losses were extensive. There were many similar other reports on this battle, and it was clear that the Japanese had used the jungle to outmanoeuvre the small British Army. By May 1942, the Japanese had forced the British out of Burma on the longest retreat in British history, and they had also driven out the Chinese army. Sadly for the Burmese, they too had been forced to flee, and flocked northwards in the thousands. One unidentified man was quoted as saying in *The World at War* series, '*There were some terrible sights up there. Men were left behind and it was heartbreaking very often to see them separated from their people, wondering if they would ever meet up with them again. They were dying in the hundreds. All [the Japanese] used to do then was just to pile them up and pour petrol over them and set fire to them and that was the end of those*'. The Japanese held their heads high and marched through the deserted cities in triumph; they occupied and controlled Burma.

During the fighting, countless British soldiers had surrendered and were taken prisoner. The Japanese, however, were well known for their views regarding surrender, and their subsequent treatment of prisoners of war. The Japanese believed that a soldier should fight to the death. They were prepared to fight to the death, and it was reinforced to them that to surrender would lose all honour for themselves, their families and their nation. This view came through in their treatment of the Allied prisoners of war and they afforded them no dignity. Their treatment of the Allies who had surrendered or been captured was incredibly inhumane. From the accounts the men in my office gave me, the war might have been bad, but this was horrific.

Many of the Allied prisoners of war worked on the Burmese railway, a supply line for the Japanese military. The 258 mile railway was built in 12 months at the cost of 106,000 lives. Not surprisingly, the railway became known as death railway, and is now well documented in many first hand accounts. Of the men who died, 90,000 were forced local labourers, and 16,000 of them were Allied POWs who were made to work for 16-18hrs a day, for endless months without a break. Many British prisoners were literally starved or worked to death; others who were not as lucky were tortured to death, in some horrendous methods.

One witness account in the *Daily Mail* (London), 2010, entitled, *Monsters of the River Kwai*, told of a Japanese Sergeant who would have his men pin his victim to the ground, and force water down the prisoners throat using a hose and bucket, until there was enough to cause severe swelling of his malnourished abdomen, and then tie barbed wire around his middle, or, '*gleefully jump up and down on him*' until the poor chap died. Many other sources document all manner of tortures, in which the Japanese would 'encourage' compliance and the rapid building of the railway. It was later said that with adequate food and water as well as humane treatment, and the absence of indiscriminate torture, the Allied soldiers and local workers could have easily built the railway and more, but with the way they were treated, Allied POWs died at a rate of one man for every railway sleeper laid.

The building of this railway was later declared a war crime and not surprisingly, with that sort of a shared history, there was a great degree of ill feeling towards the Japanese from the Allies, both the prisoners and the active soldiers. When I arrived, the war may have been over, but the nation was in a right mess. The Japanese, who had dominated Burma for so long, and in so many cruel ways, had demonstrated their core value; to never surrender, now they themselves had to surrender. I will avoid going into all the reasons as to why, as there are numerous layers of complexity involved, but suffice it to say, Japan on a global scale was no longer able to win the war, and thus surrendered, even in nations like Burma where she clearly was the stronger force. From a local perspective, it was an incredible turn of events, to see the Japanese who had so powerfully ruled the country, surrender and leave the Allies in charge; and this transition would be anything but smooth.

Chapter Thirty Five
The Clean Up

BY SEPTEMBER, 1945, WHEN I was posted to Burma, the Allies had already created a new command, the South East Asia Command, and appointed Admiral Lord Mountbatten as Supreme Commander. In an interview for *The World at War*, while trying to shed light on the complexity of this theatre in the latter half of 1945, he stated, '*I suddenly found myself responsible for an enormous area of the globe…with 128 million starving and rather rebellious people who had just been liberated, with 123,000 prisoners of war and internees, many of whom were dying, whom I had to try to recover quickly, and at the very beginning I had some 700,000 Japanese soldiers, sailors and airmen, to take in surrender, disarm and put into prison camps, awaiting transportation back. Even looking at that it sounded like a big problem, but I still had no idea what I really was in for*'.

My team and I had the job of managing and sorting out some of this transitional phase. Where the allies had been prisoners of war, they were now to be released, and the Japanese themselves were to take their place. It was essentially a complete reversal that needed to happen simultaneously, and all the while, we were in someone else's country and operating with very limited resources, and the corruption on all sides was incredible. It wasn't like anything I had ever seen before and even after a few weeks of being there, I agree with Lord Mountbatten's sentiment, 'I still had no idea what I was really in for'.

When I first arrived, I noted that the most striking aspect of Burma was the Burmese people and how pleasant they were; I liked them very much. The Burmese were very placid people, and the Indians were much sharper business people. You had to

Ali Achbar, Cryil's Burmese bearer.

really watch the shrewd Indians, but Burmese weren't silly either. In fact, most of the Burmese were very well educated, and a few of them were even petty crooks, but overall they were the nicest people I had ever met. They were always very kind and simple, it wasn't that they were uneducated; far from it, it was just that they actually valued the 'simple life' and were not as competitive as their Indian and Chinese neighbours.

I had a bearer, Ali Achbar, a local Burmese man who used to look after me. Ali would make my bed and bring me tea and do all that sort of thing. He was very good indeed. I soon discovered that he had a university degree, he was a smart chap, just not a particularly driven man. Making money and doing all that sort of thing didn't appeal to him. It was just a different set of lifestyle values he and many of the Burmese pursued. I thought, 'The best of luck to him' and couldn't help but notice that he didn't seem anywhere near as stressed as most of the British men.

Ali was kind and always looked out for me. I woke up one morning to the sound of him coming in to bring me my cup of tea. I was under my mosquito net and everything seemed normal. Suddenly, my boot flew across the room and there was a loud thud. Startled I turned to the door and saw Ali who had just thrown it. Apparently, there had been a snake, a rather poisonous one, on the other side of the mosquito net with its head up and looking at where I slept. Normally I would wait until the bearer had come in and put my tea on the dresser before I would lift the net on that side of the bed and slide my bare feet on to the floor, to sit up and drink it.

Ali told me to stay under the net and promptly left the room. He returned a short while later with two long sticks. In the ultimate display of dexterity, he used them like chopsticks, to pick up the snake, holding it firmly just behind the head and put it in a bag. I was very grateful to him, and so should the snake have been, because if I had been left to my own devices, I would have chopped it up with a garden fork! After his display of bravery, he explained that it was a krait, and that was why he felt the sticks were necessary, because just picking it up by the neck would have been too risky!

I learnt during many of the discussions that followed my

exciting morning, that drop for drop, its venom is the deadliest of any land serpents, except for a few types that are found in Australia. In fact, it was commonly believed that one bite contained enough venom to kill a dozen men and the Americans called it the 'two-step' snake, claiming that if it bit you, then that is how quickly you would die. I am sure this last part is a bit of an exaggeration, as tends to be the case with both American and snake stories, but nonetheless, I did feel quite obliged to Ali. It wasn't a particularly nice encounter; come to think of it, the only snakes I ever met were unpleasant.

Front aspect of the mess and offices in Rangoon that the RAF used as a part of their clean up operation. This building had previously been used by the Japanese as a part of their interviewing and information gathering processes, which were extremely barbaric, as evidenced by the tools and implements they left behind.

My friend, Nick, also had a bearer, who was efficient and respectable. With the war having just ended, this bearer wanted to take leave and go to visit his family in India. Nick arranged to fly him to Calcutta. To our great surprise, the bearer never returned. We never heard from him again, which was unlike him, as he was such a loyal chap and this was completely out of character. Then we got news, somehow, that he was missing. He had never actually arrived home to see his family.

Nick felt he owed some loyalty to his bearer, so he took leave and went to India to look for the young man himself. The bearer had caught the train to go to his home town, which was some way off, but he never arrived home. Nick ended up following in his footsteps and trying to trace his path. Eventually he came to the conclusion that somewhere along the train trip this bearer had disappeared, and most probably been murdered. It was figured that the man was taking a few things home to give to his family, and probably some money, and so this would have made him a target. Sadly, the local police were no help to Nick.

Nick was faithful to his servant, which was nice. We did get attached to our bearers. It is the way the world was, a loyalty existed that worked two ways. Nick was very good, he spent a lot of money trying to investigate it, and he used up all of his annual leave. Before he left India he went and visited the family, and offered to help in any way he could. He did all the right things, but he didn't stand out at the time as being particularly noble; it was just the way the world was, and even after the war, the morals of most individuals had not changed greatly. They were the values, or at least the values that we like to remember, and people were pretty loyal to each other, irrespective of their position.

Cyril sitting at his office desk in Rangoon.

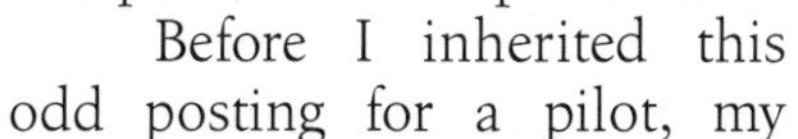

Before I inherited this odd posting for a pilot, my predecessors had needed to use large amounts of money to effectively bribe the locals as the country's economy worked on a 'cash only' bribery system. All this money was in local currency and in very low denominations; it was money that was specifically set-aside for the allied officials to use, as they needed. The money was used for things like bribing locals to sabotage the Japanese, and all sorts of necessary military tasks. Sometimes it was used for getting a person home, particularly if an aircrew had been shot down. It could be used for buying anything from train tickets to obtaining forged passports or civilian airline tickets. The cash was for anything that the forces needed in order to do business in such corrupt conditions. It was how trade was done. As a result of the events of recent years, the

The streets of Rangoon outside Cyril's office.
Note the curfew sign in the background.

entire economy was based on instantaneous cash bribes, there was never any official business, and I don't recall ever seeing a bank.

As a part of the clean up, I was given the task of collecting and gathering the money that had not been spent. There was a lot of it, and it was distributed amongst all the senior officials, all over the country. There were records of it all, so it was easy enough for it to be collected. What I hadn't counted on, was that these orders had come from someone who did not know what denominations the notes were in. It was very simple for a bloke sitting at a desk miles away to say that someone needed to collect all the money. On paper it may have seemed like a great idea, but before long we had a giant bag full of the notes. It was a great pile of money. In fact, by the time my staff and I had collected it all, it filled up one of the biggest suitcases we could find. Apart from being a physically large amount, its value amounted to the equivalent of many thousands of pounds, but all in the smallest few denominations of a local currency.

I contacted the Wing Commander and explained that we

now had such an enormous stash of money that we had no idea what else to do other than store it, in a leather case, under my bed. He realised, as did my bearer and I, that this was a serious security threat, especially in a country that was so torn apart by the war, and with civil authorities virtually not in operation. There were thieves everywhere. My bearer was very professional and could be completely trusted. He knew it was there, after all he made my bed every day, but he wasn't interested in taking any of it. He was a very honest chap, and very straight forward, and he kept telling me every day that I needed to get rid of it as fast as possible. Ali was concerned for my safety, and ever so relieved when I told him it was going.

The chief had instructed me to take the money to Headquarters, which happened to be in Calcutta, India, of all places. It was quite a task. There was a great stack of the stuff and just physically moving the case was enough for my back to remind me of its unhappy state. It was the sort of thing that you couldn't send by the postal service, even if there had been one. We couldn't take it to a bank, as these weren't operational either and burning it and just writing it off the books was apparently not the thing to do. The money had been nothing to do with me, the only part I had played was following orders to collect it. Now it was my responsibility to shift it from under my bed, and get it to India!

Working for the RAF it was easy enough to find someone who could fly me to Calcutta. My goodness, I had never been to any place in the world that was like it. There I was, getting off the aeroplane, and about to negotiate some of the densest traffic in the world, and I was hauling this oversized suitcase, stuffed full of money. Of course, no one would have known what was in my bag, but I did, and I felt very exposed, especially as I saw all these desperately poor people on the streets.

India was really something different. There were all these people literally sleeping in the streets, some of whom I had to step over to make my way down the path. Then, dispersed amongst them were some of the wealthiest people and most extravagant houses. I didn't like seeing the extremes, it just didn't feel right to me, but I particularly didn't like carrying so much money through the streets, whilst travelling on my own. It was such a great relief when I found the correct recipients and offloaded my burden.

As it happened, it was a good thing I did get rid of the money, because a few weeks after that, there was an incident, back in Burma. Whilst sleeping in my room, in the middle of the night, there came a dacoit, one of the local thieves, who would climb into a window and steal what they could. This thief picked on the room of a chap on the second floor, three doors down from mine, who himself was a bit round the bend. The first I knew of this, was that a shot was fired which woke me up. I thought maybe he was drunk again, so I quickly rushed down to investigate, along with several others.

We went into the man's room, and he shouted, 'No, no, no! Someone broke in,' and I said, 'Well what happened?' He said, 'Oh, well, I shot him'. I asked, 'Where is the body?' Apparently, this crook had climbed up the downpipe and attempted to climb in through the window, and as he was half-way in, the occupant of the room had picked up his gun and shot him. I exclaimed, 'Well, you can't do that!' but he had. We quickly got a search party together and went down, and sure enough, we found a dead body in the garden.

It took some explaining, it really did! I couldn't believe it myself, and I wondered how I was meant to explain that to anyone in authority. I went and saw my local boss and let him know what had just happened. I told him that we had a body and it was covered up with some vegetation, but it was in the garden so we needed to act quickly and move it. With the aid of various people, we managed to get rid of the body, but there was little else we could do. No one knew the victim or anything about him. The civil authority wasn't operating, and there was no one we could tell, mercifully, because there was no excuse for this shooting.

The situation was absurd! A bang like that in the middle of the night certainly does wake you up. Then to see a mad bloke running around saying, 'I shot him!' was really something else. The shooter didn't lose his job, only because there was really no authority, or no police that we could talk to and my boss, and whoever he reported to, must have thought it was a case of the man just doing what he thought was right at the time. The next day, Ali, my bearer, seemed to know all about it from the local gossip, and he was very glad that I had gotten rid of the cash from under my bed, he had always been concerned about my safety.

Apart from dealing with some of the local corruptions, my main task was to sort out the situation with the Prisoners of War. This enormous task was left to the Air Force because the locations of the prisons were so remote and difficult to access through the jungle tracks in the monsoon rains, it was usually necessary for Dakotas and other light aircraft to ferry people around. It wasn't simple for the army to just drive across land and ferry them back home, in fact in some cases, an overland route was next to impossible. Our task was two fold: to bring the allied prisoners of war home and to more or less put the Japanese in those same prisons, and then to take the Japanese out every day to the forced labour sites.

Cyril and colleague inspecting a temporary Japanese transit camp.

Since the Japanese had surrendered, there was suddenly a lot of change in many areas that needed to happen, almost simultaneously. It was thought that by forcing the Japanese to do labour, it would assist with the enormous rebuilding tasks, and it would keep them out of trouble. It was all very complicated, because suddenly, after the Japanese had occupied the land, the Allies were responsible for governing them as well, and there were millions of people to organise.

Temporary Japanese transit camp in Rangoon.

In the main cities, the labor camps were organised quickly, and we would frequently see trucks going past with full loads of Japanese men being forced to go out to work and rebuild the docks and main services. The Japanese would pass us and bow and scrape and salute us and do everything to show respect, because they were truly terrified of us. They most likely were

Photo Cyril took of Japanese soldiers working on a forced Labour camp.

imagining what they would have done, and did do, when any Allies surrendered or were taken prisoner, and now they felt that they had great reason to fear. All things considered, we did treat the Japanese prisoners pretty well, and when I would go for walks and check up on what was happening around the town, it always stood out to me that at least us Allies afforded the Japanese prisoners the dignity of enough food, clothes and even shoes.

Since the Japanese had surrendered, the allied POWs were technically no longer prisoners. Despite this, many were still living in the prisons, particularly the more remote ones, where the Air Force had not yet managed to retrieve them. They were just living there, with open gates, using the place for accommodation of sorts. The Japanese guards were no longer there and most of them were now forced labourers, or worse, but they certainly weren't guarding the Allies anymore. Some of these POW camps were a very long way from anywhere. In fact, when it comes to Burma, even the capital, Rangoon is a long way from anywhere. These men were free, and could walk out of the prisons, but there was nowhere for them to go. There was an enormous backlog of people waiting to be flown out especially from the remote prisons. Many of the prisons were not near a landing strip, so we also had to arrange for them to be transported to an airfield before they could be flown out. It was a logistics nightmare, as there were such incredible numbers of people involved, it took the better part of a year to try to sort it out and retrieve most of our men.

What we, as the RAF, did at first was to drop loads of supplies over the prisons to keep the men going while they waited to be flown out. Now that the war was over, they asked for all sorts of things that you wouldn't expect. They wanted basics like clean water, food, soap and medical supplies, and the sorts of things that they had been denied as POW's, but they also asked for other things too, things that one may classify as 'art' or 'entertainment' of sorts. I suppose they had been locked away for a long time, but still, our staff in the office did receive some very unorthodox requests for 'supplies'.

We had to organise for the packages to be dropped from the Dakotas with a parachute attached to break the fall. Apparently the medical supplies and food were particularly well packed. We obliged them with their other unusual requests and certainly had no complaints after the deliveries. Meanwhile, they apparently met a lot of the locals, although I am not sure how they went with the language barriers. I expect there may have been a sort of exchange system for worldly goods, but the depth of the relationships formed may not have been based on a lot more than that.

The situation was so bizarre, even my staff took a bit of getting used to the fact that the roles and authority had now been reversed. Not only was it challenging to move all those Allies out of prisons and then put the Japanese into the prisons or labour camps, given the vast distances and the levels of corruption and infighting going on, but I soon found there were other things I had to oversee as well. Most of the RAF aircraft based in Burma were also to be used as necessary in other nations such as Thailand and Indonesia, and there was a great demand for their services down that way too. There were a lot of interrelating and difficult resource management issues developing.

Unfortunately, there were also some challenging local situations simmering. The Burmese had long standing internal feuds that dated well before World War II. Now that the Japanese were no longer a common enemy, some of the local Burmese, who were now armed, resumed fighting their old disputes, amongst themselves. There were many little rebel groups and armies that were emerging. In fact, it brought the country further to its knees. This made travelling across the countryside by land even more treacherous. So, we really did have to fly the POW's out, which highlighted another major problem.

A new private enterprise had started. Some of the Air Force men were using service aircraft for private commercial gains. These men were making money on the side by agreeing to take civilians, or locals—often from various rebel groups, from place to place.

To make matters worse, to cover their tracks, they had to cook the books, and this meant getting people higher up the command chain to be in agreement. It took my office staff and me a little while to figure out what was going on, but when we finally did, I understood yet another reason why getting the men out of prison, and flying them back home, was taking so long. It was a very sensitive situation and it was disappointing for me. After all these years of service, to find out we ourselves were not free of corruption was pretty hard to take. Their disloyalty had hurt me personally, considering the price we had all paid with our lives for the war.

As the Intelligence Officer, part of my job was to report to the Air Officer Commander. When I realised what was going on, I went and told him, but worst of all, I had to inform him that if we cracked down on this corruption, then he would lose some of his top Squadron Commanders. The corruption was widespread and deep. The Air Officer Commander said we couldn't afford to loose so many people, so he decided to pick out a few to make examples of. It was very awkward for me to have to tell him this, but it must have been much more difficult for him to have to respond to the situation.

A typical Rangoon bus.

Burma was a lovely country and I had a fascinating time. The corruption made life difficult, but apart from that, the people had a gentle nature and were quite ingenious. In terms of resourcefulness, I particularly remember that they had many old buses. The buses broke down often, but it was just a part of their normal transport challenges. I often wondered if it was not the repository for all the buses and vehicles that had broken down in Europe and weren't quite bad enough to be scrapped. One bus pulled up and its headlights no longer worked; to resolve the problem, sitting on the dashboard were two lit candles, just in the bottom corners of the windscreen with a reflective foil plate behind them. These were the headlights. The driver couldn't use them to see anything on the road, but at least the bus could be seen by the oncoming vehicles; I didn't board that bus.

After I had been in Burma for nine months, the Group Headquarters moved to Singapore, and so I was sent with them. Why I was sent there, I am not sure, other than that the higher authorities decided that I should go. When I arrived, I found that there was nothing that I needed to do, but it was a nice place to do nothing. After a few weeks, I was told that I was free to get a ship back to the United Kingdom and just carry on with my life. I wasn't exactly sure what that meant. The one thing I knew was that I was returning home to a very different Europe.

A few of our Group boarded a ship as soon as we could. It was a large, all welded metal liberty ship. Many hours passed with me sitting on the deck, and thinking about what was in front of me now. I missed Bet greatly, and I couldn't wait to return to Liverpool, and then make my way up to Scotland, although I must confess I was a little nervous, and hoped she felt as excited as I did. From my seat at the stern of the ship I could see across the deck, all the way up to the bow. When we sailed through the Red Sea, the noonday sun was incredibly hot. One afternoon the bow of the ship slowly disappeared before us. I commented to my colleague that this was very interesting, and we realised that the ship was getting so hot under the sun the metal would slightly warp. After we moved to somewhere cooler, I spent the rest of the afternoon contemplating the same thing I had all morning; what the future would have in store and how strange it felt to be going 'home' now that it was all 'finished'. For an ending which we had all looked forward to, and paid a great price fighting for, it suddenly all felt very surreal and anticlimatic. I was no longer needed. I wasn't sure how I felt about that.

When I arrived back in the UK, I was sent to a demobilising station where they gave me a clean civilian shirt and a new suit. They even gave me a hat, pair of socks, and some new shoes. It was all part of the deal. This was pretty good of them, as clothing was still only obtained with ration coupons, so it would have been much harder for me to buy it myself. I stepped out of the fitters, and onto the streets; suddenly I was a civilian. We all joked about how much the ex-service men stood out. There we were in these brand new suits, amongst ordinary civilians; it really did look a bit odd. The next stop was home, to see my mother, but this time it felt peculiar. In a way, going home was like leaving home; I was leaving behind all the things that I had become familiar with.

Chapter Thirty Six
Home from War

I WAS STILL ENGAGED TO Miss Elizabeth Clarke, and we had continued writing letters to each other every week. Bet had been enjoying the freedoms of civilian life for a lot longer, so she had already made great strides towards 'getting on with normal life'. When I was dating her in Prestwick, she had been a Theatre Sister, but after the war in Europe ended, an incredible opportunity presented itself to her. Bet had applied for a job as an airhostess, and in those days an airhostess had to have medical training and speak at least two languages; with her war experiences, and fluency in French, she landed the job easily.

It turned out that Bet's mother had family connections, and with the Duke of Hamilton, no less! In those days, there was a labor government in power in the UK, and so it was decided that only one state run airline was to be established in England. Scottish Aviation, a company that had been growing throughout the war and manufacturing a large amount of aircraft, decided that they wanted to run their own passenger airline. Despite several applications for approval to operate out of the UK, the labor government had refused. The Duke of Hamilton in fact owned Scottish Airlines, and I rather think the fact that he was an aristocrat didn't help his case with the labor government either. When all options had been exhausted and it was clear that Scottish Aviation could not establish itself in this market, the Duke took his business elsewhere. He established three other airlines—the Royal Hellenic Airlines based in Greece, Belgium Airlines and a smaller airline, based in Luxemburg.

Bet must have been pretty terrific at her job as an airhostess, because she was sent to Europe, specifically Greece, to interview,

Scottish Airlines aircraft on the tarmac at Prestwick Airport, where Elizabeth was first employed by the company.

select and train other new applicants as airhostess for the new Hellenic airlines. She was having a great time. When the Duke came across to see how his new airline was developing, she had a particularly fantastic time! She and the Duke hired a car, and they went travelling. They visited France, Holland, Belgium,

Elizabeth in her uniform, talking with the ground crew, while working on a Luxemburg Airlines aircraft.

Switzerland and everywhere else, driving right across the continent. They were allegedly travelling to set up these airlines, but I think it was all a bit of an excuse to have a great time.

Occasionally, she would remember me, her fiancé, and write to let me know how much of a great time she was having. I was truly glad for her, and I think it was the most sensible thing I could do, to let her go, and make the most of such a fantastic opportunity. She was living the high life, and the Duke certainly travelled in style. It was a very different life to the one I was leading. I was living at home with my mother, on tight rations, constantly troubled by back pain and trying to study umpteen

Elizabeth working as a hostess for Scottish Aviation, serving delicate patisserie sweets.

hours a day. I was really glad for Bet, and thankful that she didn't have to see me struggle so much through that year. It really was a tough year for me. Nonetheless, I was sure of my love for her, and she must have been too, because we decided we would set the wedding date for immediately after I had finished my university studies.

I made the necessary arrangements to do the last year of my Engineering degree at the Wigan and District Mining and

Aida Monica Cooke, Cyril's Mother, in her kitchen apron, holding one of her rabbits.

Technical College. Several years had lapsed, but nonetheless, someone official determined that I should just recommence from where I had left off. That was probably an easier instruction for the administration officer to issue, than it was for me to actually fulfill; I had forgotten a lot, and the pace of the final year course was difficult even for those who did not have a six year gap in the middle of their program.

It felt strange and almost irrelevant to spend so much time on academic work, after all the experiences I had been through. There was however, an advantage with my situation; I now felt much more grown up and I had also learnt how to really focus and concentrate, but still, the study was rather difficult with so many of the foundation blocks either rusty or long forgotten. It was difficult to catch up and keep up with my peers, and all my study felt urgent, if not still strangely irrelevant. There weren't any allowances made either because I was one of only a handful of people in this position. Most of those who had commenced the course in the same year that I did, had found other priorities for their lives after the war, so we were a very small group of returned servicemen, straining our brains to recall basic calculus and formulas, somewhere below the layers of memories and interruptions since 1939.

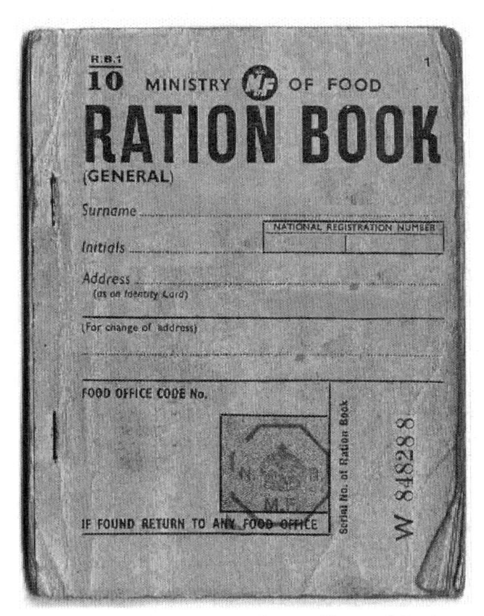
R.B.1
10 MINISTRY OF FOOD
RATION BOOK
(GENERAL)
Surname
NATIONAL REGISTRATION NUMBER
Initials
Address
(as on Identity Card)
(For change of address)
FOOD OFFICE CODE No.
IF FOUND RETURN TO ANY FOOD OFFICE
Serial No. of Ration Book
W 848288

Ration Coupon Books, still used long after the war was over.

The other students in my year were quite a bit younger—boys who had completed the first few years of university immediately after they finished school. The universities had continued running such courses as engineering all throughout the war, because there was a very real need for people to be trained in these professions. These students had been granted a deferment from their service time in the armed forces so that they could complete their studies, as it made more sense to first educate the young lads before bringing them into the services.

Meanwhile, my scholarship to undertake postgraduate studies in physics at Cambridge, had lapsed, presumably being given to one of the many young lads who kept climbing the schooling ranks, while I was away. This was disappointing but at least the war had given me the opportunity to fulfill some of my childhood dreams: becoming a pilot, sailing to Africa and exploring its legendary culture. Now, it was clear that engineering was my next challenge and the path that I should take. At this stage however there seemed to be more hard work in front of me than any of the thrills and excitements that I had anticipated when I signed up for the course ten years earlier. It would be a long time until I would be out there again, making a difference and feeling excited in my occupation, but I could see that the day would come, when all the hard work and study would be worth it.

British Propaganda poster, issued by the Ministry of Food.

It was also a particularly challenging time because my back had become increasingly painful, and now I had the time to notice it. I quickly discovered that standing at the rear of the lecture theatre was the way to go, because if I sat down for too long and settled into a chair, my back tended to seize up and it would take me a long time to be able to stand again. It was difficult, having to write all my notes out by hand, while standing up, but I got through it and whenever I dared to feel down about my situation, I reminded myself of how fortunate I was. A brief reflection on the fate of four of the other six members of my Bomber Command crew, made me feel like the luckiest student in the college.

I spent a lot of time that year taking long gentle walks around my old village. In part this was because I thought it would help my back, the pain of which had now completely caught up with me, and in part because I was now the new owner of a little dog. She was a 'bitsa', with bits of this and bits of that all interbred and I had come to own her after she had been accused of menacing the neighbor's sheep. I thought she was likely innocent, but nonetheless I was glad to have such a simple and playful companion. She could be trained to do little tricks, like pick up pieces of scrunched paper and put them into the waste paper basket. As we walked, I noticed her gentle and loyal ways—an obvious contrast to the world politics and human nature I had been contemplating.

Mrs Clarke and Mrs Johnson, the two widowed mothers of the bride and groom.

Walking gave me time to ponder and to try to make some sense of the world. While I had been in Burma, Bet had kept me up to date with the latest political developments, and I was all too aware that it was a rapidly changing Europe. In addition, we had received a lot of news in the RAF offices there, so I knew I would be coming home to a markedly different place. It seemed to me that the Allies had struggled as much with what to do after they won the war as the Germans had after losing it. Everything was in turmoil. Coming out of a war, and resuming 'normal life' was no easy business, neither on an individual level nor on an international one.

At the same time, amidst a landscape torn apart by war, and many grieving families, there were also plenty of wonderful events during that initial post-war period, and most people in the community seemed to be in good spirits, with more engagement, wedding and pregnancy announcements than we had ever experienced before. If I still had my little Wigan Newspaper it would have been filled with happy declarations and official announcements. I had my own exciting news too; I was now more in love with Bet than ever, and my dreams of living happily ever after with her, were beginning to take shape. In April 1948, after I had finished my studies, and Bet had returned from her exciting travels abroad, we counted down the days to our wedding.

The Bridal Party. Note the three bridesmaids in their traditional white dresses, with Moira Cooke left, Bet's cousin centre, Cyril's sister Elizabeth on the right. Two of the groomsmen were Cyril's friends from his Engineering degree, and the third was one of Bet's cousins.

As it turned out, weddings, particularly traditional Scottish ones, are quite a big deal. I don't think any man is ever as prepared as a woman tends to be; they must grow up dreaming about these sorts of events.

I myself would have preferred it to be a small and private gathering, but that was not to be the case. I don't think the organisation was too much work for my best man, or myself, but

The Bride and Groom—Mr and Mrs Johnson.

there were some frantic ladies involved. Us men just went along and did what we were told.

Bet's family connections extended further than I had imagined. The Clarke family had lived in that part of Scotland for centuries, literally. It turned out, that they were incredibly well connected, and it seemed as though nobody was going to miss this wedding. The wedding was rather grand, and held in one of the ancient Churches of Scotland in Kelvinside, Glasgow, and then followed by a posh reception. It was amazing how much her family and friends had pulled together, as a community, to make the day possible. Food and clothing were still rationed, so we knew that many families had made sacrifices in order to provide such grand catering. Bet's dress was marvelous, and I can't imagine how many ration coupons must have gone into obtaining the fabric. She did it! She had organised a grand wedding, but it must have been difficult given the limited availability of nearly everything.

After all the counting down, and excitement in the lead up, when the actual day came, I was terrified. I had never before seen so many people lined up to greet me. I also remember being shocked at how many female relations were there, who had lost either their fiancés or husbands to both World War I and II. It was quite startling.

After the reception, we were glad to leave the crowds and go away to a hotel where no one would know us. The next morning, we were having breakfast in the restaurant, and people kept congratulating us. I was a bit perplexed. I didn't think we should know any of them, so how did they know that we just got married? Then someone showed us the day's newspaper. Our wedding was on the front page of the *Glasgow Herald*, along with our photograph.

Chapter Thirty Seven
Life after the War

We had a lot of good news to celebrate during the rebuilding season that followed the war. I had just finished my Civil Engineering degree, when my friend and classmate, Stan Wright, received a job offer. Being in the same field, I decided to go with him to learn about the company he was going to work for; it was Simon-Carves, the fourth largest engineering firm in

Stan Wright

Cyril Johnson

Europe at the time. They mostly built power stations, so after the war, with the infrastructure repairs that were needed, they had their work cut out for them. While I was with Stan, just tagging along to see what the company was like, Stan's new boss offered me a job on the spot. I hadn't even asked. The offer just emerged out of a conversation about how Stan and I had been studying together. I gratefully said, 'Yes,' and was asked to start a few days later.

It was a fantastic company to work for. Lord Simon ran it, and he did a superb job at respecting and in turn, demanding respect from, his staff. It was a company with the traditional values that we once had, and certainly our parents had. It was the sort of company where employees would remain for their entire career. Lord Simon was very generous, and would pay for his staff to do night courses and all sorts of other educational programs. If we needed to leave early to attend classes or anything similar, this was acceptable. He was always very generous, and appreciated that it was important to build up his staff and in return, we were trusted to manage our own hours and contribute what was necessary. It worked! We all worked, and hard! There was mutual respect, and people were very loyal to one another and to the company. Stan had a girlfriend, Julie, who also worked for the firm. Stan and Julie eventually married, and we kept in touch, with Julie still writing to me more than 60 years after we met.

Photo of Aida Monica Cooke just before she emigrated to Australia, in her newly-obtained British Passport.

Early in 1949 my mother left for Australia. My uncle Oswald, 'Aussie' Cook, and his wife had been living in Australia for decades, teaching Cobb and Co's drivers to drive busses, but Aussie's wife had recently died. Soon after, he invited my mother,

The S.S. *Stratheden*, bound for Australia, 3rd July 1959.

who was by then in her seventies, and my sister, Elizabeth, to come to Australia to keep him company. My mother thought this was a wonderful idea and a short time later she and my sister emigrated to Australia.

We had been married two years when Elizabeth and I had our first son. We were so delighted; life was wonderful. A year later, while I was at work one day, just minding my own business, the general manager at Simon-Caves came and said, 'Would you like to go to Australia?' I said, 'Not particularly, I have never thought about it'. The manager pressed on, 'Well, doesn't your mother live there?' I replied, 'Well, yes, but so?' as he pressed on, 'Well, wouldn't you like to see her'. I said hesitantly, 'Okay', still unsure where this conversation was going. He quickly replied, 'Oh good, because we have a ticket booked for you and you are to leave next week'. Before I had a chance to ask about my wife and small child, he offered, 'The company will pay for someone to pack up the house and see to it that Bet and your son will be sent across on a ship'—which they did, in two first class cabins. They really did look after us.

Australia was a great place to live, but after the project

Cyril and Elizabeth Johnson boarding the S.S. *Stratheden* bound for Australia.

with Simon-Calves finished, we returned to England in 1953. A few years later, we had a second son, and not long after that, a daughter. We were a very happy family. One cold winters day, with a house full of children stuck inside, Bet suggested that we apply for another job in Australia. We had enjoyed the Australian climate and the people were so easy to get on with. I applied for and landed an engineering job in Adelaide, South Australia. On 3rd July, 1959 we boarded the S.S. *Stratheden*, and moved to Australia. As can be seen by our photos, we were particularly excited about our new lives. We could hardly contain ourselves.

Bet and I enjoyed a happy and successful marriage, and we were very proud of our three children. My days with her were the happiest of my life. Of course, we had our share of hard times, everybody does, but we found that through our trials our marriage got stronger and stronger. As we grew older together, our love continued to grow deeper; those last years with her were like a glorious sunset. We walked side by side throughout life, sharing every turn of events, for richer and certainly for poorer, in sickness and in health; she put up with a lot. No matter, we were constantly by each other's side for just a few weeks shy of 50 years. Our secret? The proverb Bet used to always quote, 'never let the sun go down on a quarrel'.

Cyril and Elizabeth Johnson participating in the ship's entertainment, with Elizabeth dressed up as the luggage, riding on her son's scooter.

Afterword

From the Author

CYRIL HAS JUST CELEBRATED HIS 94th birthday in January 2014, and still lives independently in Adelaide. He has three children, eight grand children and a great-grandchild, and is well supported by his close family, many of whom have inherited various aspects of Cyril's character, passion and interests.

Cyril's eldest son, Alastair, is a man of strong character with great determination and discipline, who trained with the Adelaide University Squadron and was commissioned as a Pilot Officer in the Royal Australian Air Force, before embarking on his own career path. In the next generation, Alastair's eldest son, Andrew is currently a Captain in the Australian Army, and displays many of the leadership characteristics of his father, grandfather and great-grandfather.

Cyril's second son, Steve, my father, has an unwavering passion for aviation and civil engineering. Steve built model airplanes with Cyril from a young age, and went on to become a private pilot, who currently enjoys flying in his own Tiger Moths. In an uncanny way, Steve's son, Ryan, shares the same passion, and is a qualified commercial pilot. It is an incredible sight to see the men from all three generations go out into our fields and start up one of Steve's vintage 1939 Tiger Moths and help Cyril to climb up into it, and take him out to enjoy some aerobatics over the beach.

Yet the similarities, life interests and passions have not stopped there. As Cyril's career path developed, he undertook many significant engineering projects and was frequently in project management positions requiring leadership and ingenuity. A generation later, Steve has also become a Project

Director and Fellow of the Institute of Civil Engineers. He is still experiencing an extraordinary career path, which is in many ways similar to that of Cyril's, and the careers of both men would be subject enough for an entire book, each.

Furthermore, Cyril's father was profoundly impacted by his experiences in Africa, and this came through in the stories that he passed onto his son. Just as Cyril went on to have many amazing adventures of his own, his son Steve also travelled extensively with his daughter, Jaclyn, exploring Africa where they both had many life changing experiences.

Additionally, when Cyril was a child, he lived on a property where his grandfather and two uncles ran the local doctors' surgery, and there were a great number of doctors in the family tree. Within the family there was always a great respect for medical professionals, so it was no surprise when Cyril's daughter, Trish, became a General Practitioner. With Cyril's debilitating back pain, malaria relapses, deafness and the ongoing repercussions from his childhood rheumatic fever, this was indeed a very useful set of skills.

In the next generation, I myself am taking up the family tradition and am currently studying medicine.

If you asked all of these people, I know they would tell you that they have their own reasons for their chosen paths in life, and yet, when the similarities are viewed from an aerial perspective, the themes are uncanny. So what does all this mean? As I observed the repeating themes, it became abundantly clear that some of these characteristics, interests and passions had been passed down for multiple generations. To me, this shows unmistakably, that the lives we live have meaning and significance for those who are yet to follow. Furthermore, in an age where people are interested in self-discovery, perhaps the first place to start is by honouring their previous generations and taking the time to understand who they really were.

I would sincerely encourage everyone with a parent or grandparent who has lived through such fascinating times to make the time to hear and record their stories in a way that can be passed down to future generations. They have doubtless shaped you in more ways than you are aware.

Bibliography

Books:

Chant, C. (2003) *Lancaster: The History of Britain's Most Famous World War II Bomber*. Bath. Parragon.

Hastings, M. (1979) *RAF Bomber Command*. Minneapolis, Zenith Press.

Hitler, A., & Manheim, R. (1971) *Mein Kampf*. Boston: Houghton Mifflin.

Iveson, T. (2009) *Lancaster: The Biography*. London. Andre Deutsch.

Kapuscinski, R. (1998) *The Shadow of the Sun*. London. Penguin.

McKay, S. (2010) *The Secret Life of Bletchley Park: The WWII Codebreaking Centre and the Men and Women who Worked There.* London. Aurum Press.

McKinstry, L. (2009) *Lancaster: the Second World War's Greatest Bomber*. London. John Murray.

Owen, T. (1978) *Aircraft of the Royal Air Force Since 1918.* 7th edition. London. Macdonald & Jane's.

Richards, D. (1953) *The Royal Air Force 1939–1945 Volume 1: The Fight at Odds*. London. Her Majesty's Stationery Office.

Saward (1984) *'Bomber' Harris, the Authorised Biography.* Bath. Cassell Buchan and Enright.

Stewart, O. (1941) *The Royal Air Force in Pictures including Aircraft of the Fleet Arm.* London. Country Life Limited.

Taylor, J. (1969) *"Avro Lancaster" Combat Aircraft of the World from 1909 to the Present.* New York. G.P. Putnam's Sons.

Tedder, A. (1966) *With Prejudice: The War Memoirs of Marshal of the Royal Air Force, Lord Tedder*. London. Cassell.

Online:

Royal Air Force Ministry of Defence, 576 Squadron.
Available at:
http://www.raf.mod.uk/history/bombercommandno576squadron.cf
accessed on 1st December 2013.

The Hindu Newspaper.
Available at:
http://www.thehindu.com/todays-paper/tp-international/nigeria-dye-pits-kept-alive-by-local-techniques-despite-threats/article5184271.ece
accessed 1st December 2013.

Urduhart, A. Monsters of the River Kwai. *Daily Mail* London, 26th February 2010.
Available at:
http://business.highbeam.com/5900/article-1G1-219913293/monsters-river-kwai-torture-pows-burma-death-railway
accessed on 12th December 2013.

World War 2 Today.
Available at:
http://ww2today.com/3rd-december-1943-orchestrated-hell-murrow-reports-from-over-berlin
accessed on 22nd December 2013

Television Series:

Batty, P. (Writer) & Pett, J. (Director). (1973) The Desert: North Africa 1940-1943 In Isaacs, J. (Executive Producer) *The World at War*. London. Thames Television.

Williams, J. (Writer) & Pett, J. (Director). (1973) It's A Lovely Day Tomorrow: Burma 1942-1944 In Isaacs, J. (Executive Producer) *The World at War.* London. Thames Television.

Photographic acknowledgements:

Whilst the vast majority of photographs used in this book were taken by Cyril or are part of his family albums, we would like to sincerely thank those who have contributed. We are especially thankful to our relatives who have thoroughly researched the family tree and in so doing have been able to provide many images.

We are also truly grateful to the many museums, community based websites and collectors who have made available images that are now in the public domain. By very nature of these images now being in the public domain, they are often used in multiple places across the internet without citations, thus making it difficult to track their original sources. For this reason, we wish to thank all those who have contributed to the large publicly available databases of photos whose copyright restrictions have expired, often owing to the time or context in which they were taken.

Where possible, we have sought permission to use photographs as a courtesy. However, on occasion, due to photographs having been reproduced multiple times online, we have been unable to identify the original source to be able to write and obtain permission for use. In the case of any unintentional infringements please advise us and this will be rectified in subsequent print runs.

A specific thank you goes to:

Royal Australian Air Force Museum, Melbourne
Air Force Museum of New Zealand.

Bibliography

THANK YOU!